T0210475

Numerical Integration of Space Fractional Partial Differential Equations

Vol 1 - Introduction to Algorithms and Computer Coding in R

Synthesis Lectures on Mathematics and Statistics

Editor
Steven G. Krantz, *Washington University, St. Louis*

Essentials of Applied Mathematics for Engineers and Scientists, Second Edition
Robert G. Watts
2012

Chaotic Maps: Dynamics, Fractals, and Rapid Fluctuations
Goong Chen and Yu Huang
2011

Matrices in Engineering Problems
Marvin J. Tobias
2011

The Integral: A Crux for Analysis
Steven G. Krantz
2011

Statistics is Easy! Second Edition
Dennis Shasha and Manda Wilson
2010

Lectures on Financial Mathematics: Discrete Asset Pricing
Greg Anderson and Alec N. Kercheval
2010

Jordan Canonical Form: Theory and Practice
Steven H. Weintraub
2009

The Geometry of Walker Manifolds
Miguel Brozos-Vázquez, Eduardo García-Río, Peter Gilkey, Stana Nikčević, and Ramón Vázquez-Lorenzo
2009

An Introduction to Multivariable Mathematics
Leon Simon
2008

Jordan Canonical Form: Application to Differential Equations
Steven H. Weintraub
2008

Statistics is Easy!
Dennis Shasha and Manda Wilson
2008

A Gyrovector Space Approach to Hyperbolic Geometry
Abraham Albert Ungar
2008

Numerical Integration of Space Fractional Partial Differential Equations:
Vol 1 - Introduction to Algorithms and Computer Coding in R
Younes Salehi and William E. Schiesser

ISBN: 978-3-031-01283-9   paperback
ISBN: 978-3-031-02411-5   ebook

DOI 10.1007/978-3-031-02411-5

A Publication in the Springer series
SYNTHESIS LECTURES ON MATHEMATICS AND STATISTICS

Lecture #19
Series Editor: Steven G. Krantz, *Washington University, St. Louis*
Series ISSN
Print 1938-1743   Electronic 1938-1751

Numerical Integration of Space Fractional Partial Differential Equations

Vol 1 - Introduction to Algorithms and Computer Coding in R

Younes Salehi
Razi University

William E. Schiesser
Lehigh University

SYNTHESIS LECTURES ON MATHEMATICS AND STATISTICS #19

ABSTRACT

Partial differential equations (PDEs) are one of the most used widely forms of mathematics in science and engineering. PDEs can have partial derivatives with respect to (1) an initial value variable, typically time, and (2) boundary value variables, typically spatial variables. Therefore, two fractional PDEs can be considered, (1) fractional in time (TFPDEs), and (2) fractional in space (SFPDEs). The two volumes are directed to the development and use of SFPDEs, with the discussion divided as:

- Vol 1: Introduction to Algorithms and Computer Coding in R

- Vol 2: Applications from Classical Integer PDEs.

Various definitions of space fractional derivatives have been proposed. We focus on the *Caputo* derivative, with occasional reference to the *Riemann-Liouville* derivative.

The Caputo derivative is defined as a convolution integral. Thus, rather than being *local* (with a value at a particular point in space), the Caputo derivative is *non-local* (it is based on an integration in space), which is one of the reasons that it has properties not shared by integer derivatives.

A principal objective of the two volumes is to provide the reader with a set of documented R routines that are discussed in detail, and can be downloaded and executed without having to first study the details of the relevant numerical analysis and then code a set of routines.

In the first volume, the emphasis is on basic concepts of SFPDEs and the associated numerical algorithms. The presentation is not as formal mathematics, e.g., theorems and proofs. Rather, the presentation is by examples of SFPDEs, including a detailed discussion of the algorithms for computing numerical solutions to SFPDEs and a detailed explanation of the associated source code.

KEYWORDS

space fractional partial differential equations (SFPDEs), initial value (temporal) conditions, boundary value (spatial) conditions; nonlinear SFPDEs, numerical algorithms for SFPDEs, fractional calculus

Contents

Preface

Partial differential equations (PDEs) are one of the most used widely forms of mathematics in science and engineering. PDEs can have partial derivatives with respect to (1) an initial value variable, typically time, and (2) boundary value variables, typically spatial variables. Therefore, two fractional PDEs (FPDEs) can be considered, (1) fractional in time (TFPDEs), and (2) fractional in space (SFPDEs). The books[1] are directed to the development and use of SFPDEs.

FPDEs have features and solutions that go beyond the established integer PDEs (IPDEs), for example, the classical field equations including the Euler, Navier-Stokes, Maxwell and Einstein equations. FPDEs therefore offer the possibility of solutions that have features that better approximate physical/chemical/biological phenomena than IPDEs.

Fractional calculus dates back to the beginning of calculus (e.g., to Leibniz, Riemann and Liouville), but recently there has been extensive reporting of applications, typically as expressed by TFPDE/SFPDEs. In particular, SFPDEs are receiving broad attention in the research literature, especially when applied to the computer-based modeling of heterogeneous media. For example, SFPDEs are being applied to living tissue (with potential applications in biomedical engineering, biology and medicine).

Various definitions of space fractional derivatives have been proposed. Therefore, as a first step in the use of SFPDEs, a definition of the derivative must be selected. In both books, we focus on the *Caputo* derivative, with occasional reference to the *Riemann–Liouville* derivative.

The Caputo derivative has at least two important advantages:

1. For the special case of an integer derivative, the usual properties of integer calculus follow. For example, the Caputo derivative of a constant is zero.

2. The definition of a Caputo derivative is based the integral of an integer derivative. Therefore, the established algorithms for approximating integer derivatives can be used. For the numerical methods that follow, the integer derivatives are approximated with splines.

The Caputo derivative is defined as a convolution integral. Thus, rather than being *local* (with a value at a particular point in space), the Caputo derivative is *non-local*, (it is based on an integration in space), which is one of the reasons that it has properties not shared by integer derivative).

---

[1]The two volume set has the titles:
Numerical Integration of Space Fractional Partial Differential Equations
Vol 1: Introduction to Algorithms and Computer Coding in R
Vol 2: Applications from Classical Integer PDEs.

A parameter of the Caputo derivative that is of primary interest is the order of the derivative, which is fractional, with integer order as a special case. The various example applications that follow generally permit the variation of the fractional order in computer-based analysis.

The papers cited as a source of the SFPDE models generally consist of a statement of the equations followed by reported numerical solutions. Generally, little or no information is given about how the solutions were computed (the algorithms) and in all cases, the computer code that was used to calculate the solutions is not provided.

In other words, what is missing is: (1) a detailed discussion of the numerical methods used to produce the reported solutions and (2) the computer routines used to calculate the reported solutions. For the reader to complete these two steps to verify the reported solutions with reasonable effort is essentially impossible.

A principal objective of the books is therefore to provide the reader with a set of documented R routines that are discussed in detail, and can be downloaded and executed without having to first master the details of the relevant numerical analysis and then code a set of routines.

The example applications are intended as introductory and open ended. They are based mainly on classical (legacy) IPDEs. The focus in each chapter is on:

1. A statement of the SFPDE system, including initial conditions (ICs), boundary conditions (BCs) and parameters.

2. The algorithms for the calculation of numerical solutions, with particular emphasis on splines.

3. A set of R routines for the calculation of numerical solutions, including a detailed explanation of each section of the code.

4. Discussion of the numerical solution.

5. Summary and conclusions about extensions of the computer-based analysis.

In summary, the presentation is not as formal mathematics, e.g., theorems and proofs. Rather, the presentation is by examples of SFPDE applications, including the details for computing numerical solutions, particularly with documented source code. The authors would welcome comments, especially pertaining to this format and experiences with the use of the R routines. Comments and questions can be directed to wes1@lehigh.edu.

Younes Salehi and William E. Schiesser
November 2017

CHAPTER 1

Introduction to Fractional Partial Differential Equations

1.1 INTRODUCTION

Partial differential equations (PDEs) can have partial derivatives with respect to (1) an initial value variable, typically time, and (2) boundary value variables, typically spatial variables. Therefore, two fractional PDEs (FPDEs) can be considered, (1) fractional in time (TFPDEs), and (2) fractional in space (SFPDEs). The subsequent discussion pertains to SFPDEs.

As a first step in the numerical (computer-based) study of SFPDEs, algorithms for the approximation of the fractional derivatives are required. Here we consider an approach to the approximation of fractional derivatives reported in [2]. In particular, the *Caputo fractional derivative*, considered next, is defined as

$$\frac{d^{\alpha}u(x)}{dx^{\alpha}} = \frac{1}{\Gamma(n-\alpha)} \int_a^x \frac{d^n u(\xi)}{d\xi^n}(x-\xi)^{n-\alpha-1}d\xi \tag{1.1}$$

where

$u(x)$	function to be differentiated
x	independent variable, typically space for a SFPDE
a	lower limit of the integral
$\dfrac{d^{\alpha}u(x)}{dx^{\alpha}}$	fractional derivative of $u(x)$ with respect to x
α	order of the fractional derivative $1 \le \alpha < 2$
n	smallest integer greater than α
Γ	gamma function

Equation (1.1) indicates the Caputo derivative is defined in terms of a convolution integral. This definition requires $\alpha \neq n$ (from $\dfrac{1}{\Gamma(n-\alpha)}$). The integer derivative $\dfrac{d^n u(\xi)}{d\xi^n}$ can be computed with any of the established approximations for integer derivatives, e.g., finite differences (FDs), finite elements (FEs), splines.[1]

For the interval $1 \leq \alpha < 2$, $n = 2$ (from the definition of n as the smallest integer greater than α).[2] Then the integral of eq. (1.1) can be approximated as [2]

$$I_j = \frac{1}{\Gamma(2-\alpha)} \int_a^{x_j} \phi_j(\xi)(x_j - \xi)^{1-\alpha} d\xi \tag{1.2a}$$

where ϕ_j is the spline basis function at point j in x, that is, $x = a + j\Delta x$ with $j = 0, 1, ..., N$ which defines a spatial grid of $N + 1$ points in x with index j. Δx is the uniform spacing of the spatial interval $a \leq x \leq N\Delta x$.

$$\phi_j(\xi) = \sum_{k=0}^{j} \frac{\partial^2 u(x_k)}{\partial \xi^2} \phi_{j,k}(\xi) \tag{1.2b}$$

For linear splines,

$$\phi_{j,k}(\xi) = \begin{cases} \dfrac{\xi - x_{k-1}}{x_{k+1} - x_k}, & x_{k-1} \leq \xi \leq x_k \\[2mm] \dfrac{x_{k+1} - \xi}{x_k - x_{k-1}}, & x_k \leq \xi \leq x_{k+1} \\[2mm] 0, & \text{otherwise} \end{cases} \tag{1.2c}$$

(with $j = 0, 1, ..., N$). For $k = 0, k = j$,

$$\phi_{j,0}(\xi) = \begin{cases} \dfrac{x_1 - \xi}{x_1 - x_0}, & x_0 \leq \xi \leq x_1 \\[2mm] 0, & \text{otherwise} \end{cases} \tag{1.2d}$$

$$\phi_{j,j}(\xi) = \begin{cases} \dfrac{\xi - x_{j-1}}{x_j - x_{j-1}}, & x_{j-1} \leq \xi \leq x_j \\[2mm] 0, & \text{otherwise} \end{cases} \tag{1.2e}$$

---

[1]The spatial approximations discussed next are termed *splines* in [2]. Alternative terminology would be *finite elements* (FEs), specifically with *hat basis functions*.

[2]This interval in α was selected to interpolate between the orders integer **1** for first order hyperbolic (convective) PDEs and integer **2** for parabolic (diffusive) PDEs.

Substitution of eqs. (1.2c,d,e) in eqs. (1.2a,b) gives [2]

$$I_j = \frac{1}{\Gamma(2-\alpha)} \sum_{k=0}^{j} \frac{\partial^2 u(x_k)}{\partial \xi^2} \int_{a}^{x_j} \phi_j(\xi)(x_j - \xi)^{1-\alpha} d\xi$$

$$I_j = \frac{\Delta x^{2-\alpha}}{\Gamma(4-\alpha)} \sum_{k=0}^{j} \frac{\partial^2 u(x_k)}{\partial \xi^2} a_{j,k} \tag{1.2f}$$

where

$$a_{j,k} = \begin{cases} (j-1)^{3-\alpha} - j^{2-\alpha}(j-3+\alpha), & k = 0 \\[2mm] (j-k+1)^{3-\alpha} - 2(j-k)^{3-\alpha} + (j-k-1)^{3-\alpha}, & 1 \le k \le j-1 \\[2mm] 1, & k = j \end{cases} \tag{1.2g}$$

The approximation of the second derivative, $\dfrac{\partial^2 u(x_k)}{\partial \xi^2}$, in eq. (1.2f) can be accomplished with any standard method. Here low orders FDs are used.

$$u_{xx,k} = \frac{\partial^2 u(x_k)}{\partial x^2} \approx \frac{u_{k+1} - 2u_k + u_{k-1}}{\Delta x^2} \tag{1.2h}$$

To avoid a fictitious point outside the boundary at $x = x_0$ ($k = 0$), a noncentered FD is used.

$$u_{xx,0} = \frac{\partial^2 u(x_0)}{\partial x^2} \approx \frac{2u_0 - 5u_1 + 4u_2 - u_3}{\Delta x^2} \tag{1.2i}$$

Then eq. (1.2f) is

$$I_j \approx \frac{\Delta x^{2-\alpha}}{\Gamma(4-\alpha)} \left(u_{xx,0} a_{j,0} + \sum_{k=1}^{j} u_{xx,k} a_{j,k} \right) \tag{1.2j}$$

The approximation of the fractional derivative of eq. (1.2j) (i.e., $\dfrac{d^\alpha u(x)}{dx^\alpha} = I_j$ from eqs. (1.1), (1.2a), (1.2j)) is used in the numerical (computer-based) calculations that follow.

1.2   COMPUTER ROUTINES, EXAMPLE 1

A main program and subordinate routine, programmed in R,[3] based on the method of lines (MOL)[4] are discussed next.

---

[3]R is a quality open source scientific programming system that can be easily downloaded from the Internet (http://www.R-project.org/). In particular, R has (i) vector-matrix operations that facilitate the programming of linear algebra, (ii) a library of quality ordinary differential equation (ODE) integrators, and (iii) graphical utilities for the presentation of numerical ODE/PDE solutions. All of these features and utilities are demonstrated through the applications in this book.

[4]The method of lines is a general approach to the numerical integration (solution) of PDEs in which the spatial (boundary value) derivatives are approximated algebraically, typically by finite differences (FDs), finite elements (FEs), Galerkin (weighted residual) methods, spectral methods, least squares. The remaining initial value independent variable, usually time, leads to initial values ODEs that are integrated numerically, usually by a library ODE integrator.

The SFPDE is ([1], pp 1140–1142)

$$\frac{\partial u}{\partial t} = d(x)\frac{\partial^{1.5}u}{\partial x^{1.5}} + p(x,t) \tag{1.3a}$$

with an initial condition (IC)

$$u(x, t = 0) = (x^2 + 1)\sin(1) \tag{1.3b}$$

and Dirichlet boundary conditions (BCs)

$$u(x = 0, t) = \sin(t + 1); \; u(x = 1, t) = 2\sin(t + 1) \tag{1.3c,d}$$

The diffusivity and source term in eq. (1.3a) are

$$d(x) = \Gamma(1.5)x^{0.5} \tag{1.3e}$$

$$p(x, t) = (x^2 + 1)\cos(t + 1) - 2x\sin(t + 1) \tag{1.3f}$$

The analytical solution for eqs. (1.3a)–(1.3f) is

$$u_a(x, t) = (x^2 + 1)\sin(t + 1) \tag{1.3g}$$

which is used to verify the numerical solution of eqs. (1.3a)–(1.3f).

1.2.1 MAIN PROGRAM

A main program for eqs. (1.3a)–(1.3g) follows.

Listing 1.1: Main program for eqs. (1.3a)–(1.3g)

```
#
# SFPDE
#
#    ut=d(x)*(d^alpha u/dx^alpha)+p(x,t)
#
#    xl < x < xu, 0 < t < tf, xl=0, xu=1
#
#    u(x,t=0)=x^2+1*sin(1)
#
#    u(x=xl,t)=sin(t+1); u(x=xu,t)=2*sin(t+1)
#
#    d(x)=gamma(1.5)*x^0.5
#
```

```
#   p(x,t)=(x^2+1)*cos(t+1)-2*x*sin(t+1)
#
#   ua(x,t)=(x^2+1)*sin(t+1)
#
# Delete previous workspaces
  rm(list=ls(all=TRUE))
#
# Access functions for numerical solution
  library("deSolve");
  setwd("f:/fractional/sfpde/chap1");
  source("pde1a.R");
#
# Parameters
  alpha=1.5;
#
# d(x)
  d=function(x,t) gamma(1.5)*x^(0.5);
#
# p(x,t)
  p=function(x,t) (x^2+1)*cos(t+1)-2*x*sin(t+1);
#
# Analytical solution
  ua=function(x,t) (x^2+1)*sin(t+1);
#
# Initial condition function (IC)
  f=function(x) ua(x,0);

# Boundary condition functions (BCs)
  g_0=function(t) ua(xl,t);
  g_L=function(t) ua(xu,t);
#
# Spatial grid
  xl=0;xu=1;nx=11;dx=(xu-xl)/(nx-1);
  xj=seq(from=xl,to=xu,by=dx);
  cd=dx^(-alpha)/gamma(4-alpha);
#
# Independent variable for ODE integration
  t0=0;tf=1;nt=6;dt=(tf-t0)/(nt-1);
  tout=seq(from=t0,to=tf,by=dt);
```

```
#
# a_jk coefficients
  A=matrix(0,nrow=nx-2,ncol=nx-1);
  for(j in 1:(nx-2)){
    for(k in 0:j){
    if (k==0){
      A[j,k+1]=(j-1)^(3-alpha)-j^(2-alpha)*(j-3+alpha);
    } else if (1 <= k && k<=j-1){
      A[j,k+1]=(j-k+1)^(3-alpha)-2*(j-k)^(3-alpha)+(j-k-1)^(3-
          alpha);
    } else
      A[j,k+1]=1;
    }
  }
#
# Initial condition
  nx=nx-2;
  u0=rep(0,nx);
  for(j in 1:nx){
    u0[j]=f(xj[j+1]);}
  ncall=0;
#
# ODE integration
  out=lsode(y=u0,times=tout,func=pde1a,
      rtol=1e-12,atol=1e-12,maxord=5);
  nrow(out)
  ncol(out)
#
# Allocate array for u(x,t)
  nx=nx+2;
  u=matrix(0,nt,nx);
#
# u(x,t), x ne xl,xu
  for(i in 1:nt){
    for(j in 2:(nx-1)){
      u[i,j]=out[i,j];
    }
  }
#
```

```
# Reset boundary values
  for(i in 1:nt){
   u[i,1]=g_0(tout[i]);
  u[i,nx]=g_L(tout[i]);
  }
#
# Numerical, analytical solutions, maximum difference
  uap=matrix(0,nt,nx);
  for(i in 1:nt){
    for(j in 1:nx){
      uap[i,j]=ua((j-1)*dx,(i-1)*dt);
    }
  max_err=max(abs(u-uap));
  }
#
# Tabular numerical, analytical solutions,
# difference
  cat(sprintf("\n       t       x      u(x,t)
                     ua(x,t)        diff"));
  for(i in 1:nt){
  iv=seq(from=1,to=nx,by=1);
  for(j in iv){
    cat(sprintf("\n %6.2f%6.2f%10.5f%10.5f%12.3e",
      tout[i],xj[j],u[i,j],uap[i,j],u[i,j]-uap[i,j]));
  }
  cat(sprintf("\n"));
  }
#
# Plot numerical, analytical solutions
  matplot(xj,t(u),type="l",lwd=2,col="black",lty=1,
    xlab="x",ylab="u(x,t)",main="");
  matpoints(xj,t(uap),pch="o",col="black");
#
# Display maximum error
  cat(sprintf(" Maximum error = %6.2e \n",max_err));
#
# Plot error at t = tf
  err_1=abs(u[nt,]-ua(xj[1:nx],tf));
  plot(xj,err_1,type="l",xlab="x",
```

```
        ylab="Max Error at t = tf",
        main="",col="black")
#
# Calls to ODE routine
  cat(sprintf("\n\n  ncall = %3d\n",ncall));
```

We can note the following details about this main program.

- Brief comments defining the test problem are followed by the deletion of previous files.

```
#
# SFPDE
#
#   ut=d(x)*(d^alpha u/dx^alpha)+p(x,t)
#
#   xl < x < xu, 0 < t < tf, xl=0, xu=1
#
#   u(x,t=0)=x^2+1*sin(1)
#
#   u(x=xl,t)=sin(t+1); u(x=xu,t)=2*sin(t+1)
#
#   d(x)=gamma(1.5)*x^0.5
#
#   p(x,t)=(x^2+1)*cos(t+1)-2*x*sin(t+1)
#
#   ua(x,t)=(x^2+1)*sin(t+1)
#
# Delete previous workspaces
  rm(list=ls(all=TRUE))
```

- The ODE integrator library deSolve is accessed. Note that the setwd (set working directory) uses / rather than the usual \.

```
#
# Access functions for numerical solution
  library("deSolve");
  setwd("f:/fractional/sfpde/chap1");
  source("pde1a.R");
```

pde1a is the routine for the MOL/ODEs (discussed subsequently).

- The order of the fractional derivative in eq. (1.3a) is defined.

```
#
# Parameters
  alpha=1.5;
```

- Functions for $d(x)$ (eq. (1.3e)) and $p(x,t)$ (eq. (1.3f)) in eq. (1.3a) and the analytical solution of eq. (1.3g) are defined.

```
#
# d(x)
  d=function(x,t) gamma(1.5)*x^(0.5);
#
# p(x,t)
  p=function(x,t) (x^2+1)*cos(t+1)-2*x*sin(t+1);
#
# Analytical solution
  ua=function(x,t) (x^2+1)*sin(t+1);
```

- A function for IC (1.3b) is defined as the analytical solution at $t = 0$.

```
#
# Initial condition function (IC)
  f=function(x) ua(x,0);
```

- Functions for BCs (1.3c,d) are defined as the analytical solution at $x = x_l, x_u$.

```
#
# Boundary condition functions (BCs)
  g_0=function(t) ua(xl,t);
  g_L=function(t) ua(xu,t);
```

- A spatial grid of 11 points is defined for $x_l = 0 \le x \le x_u = 1$, so that xj=0,0.1,...,1.

```
#
# Spatial grid
  xl=0;xu=1;nx=11;dx=(xu-xl)/(nx-1);
  xj=seq(from=xl,to=xu,by=dx);
  cd=dx^(-alpha)/gamma(4-alpha);
```

cd is a coefficient in eq. (1.2j) that is used in the ODE routine pde1a discussed subsequently. cd is passed to pde1a without any particular specification, a feature of R.[5]

---

[5]The specification of space and time scales, particularly the number of points (e.g., nx,nout) is typically done by trial and error. The number of points in the space scale is a balance between efficient computation (an upper bound) and spatial

- An interval in t of 6 points is defined for $0 \le t \le 1$ so that tout=0,0.2,...,1.[6]

```
#
# Independent variable for ODE integration
  t0=0;tf=1;nt=6;dt=(tf-t0)/(nt-1);
  tout=seq(from=t0,to=tf,by=dt);
```

- The coefficients $a_{j,k}$ of eq. (1.2g) are defined with a series of ifs.

```
#
# a_jk coefficients
  A=matrix(0,nrow=nx-2,ncol=nx-1);
  for(j in 1:(nx-2)){
    for(k in 0:j){
    if (k==0){
      A[j,k+1]=(j-1)^(3-alpha)-j^(2-alpha)*(j-3+alpha);
    } else if (1 <= k && k<=j-1){
      A[j,k+1]=(j-k+1)^(3-alpha)-2*(j-k)^(3-alpha)+(j-k-1)^(3-alpha);
    } else
      A[j,k+1]=1;
    }
  }
```

- IC (1.3b) is defined.

```
#
# Initial condition
  nx=nx-2;
  u0=rep(0,nx);
  for(j in 1:nx){
    u0[j]=f(xj[j+1]);}
  ncall=0;
```

resolution (a lower bound). The selection of the number of points can be guided by an analytical solution if available (by considering the difference between the numerical and analytical solutions, as in this test problem) but for most applications, an analytical solution is not available. In this case, the number of points can be varied and the effect on the numerical solution observed. The accuracy of the solution is inferred by the number of reproducible leading digits as the number of points is varied (usually termed h refinement).

[6]The time scale is determined by the rate of response of the solution, and the number of points is determined by the output requirements, particularly for plotting. A selected time scale that is too short will lead to an incomplete solution. A selected time scale that is too long will give a solution with the essential variation appearing only at the beginning of time.

With nx=nx-2=11-2=9, the ICs are specified for the 9 ODEs at the interior points in x, and do not include the boundary points $x = 0, 1$ since for the latter, the dependent variables $u(x = 0, t), u(x = 1, t)$ are defined by Dirichlet BCs (1.3c,d) and not by ODEs.

The coding pertaining to nx was selected since nx is also used in the ODE/MOL routine pde1a discussed subsequently to size vectors and control fors. As explained in the subsequent discussion, the value of nx passed to pde1a is the number of ODEs, 9.

The number of calls to pde1a is initialized and passed to pde1a.

- The system of 9 MOL/ODEs is integrated by the library integrator lsode[7] (available in deSolve). As expected, the inputs to lsode are the ODE function, pde1a, the IC vector u0, and the vector of output values of t, tout. The length of u0 (e.g., 9) informs lsode how many ODEs are to be integrated. func,y,times are reserved names.

```
#
# ODE integration
  out=lsode(y=u0,times=tout,func=pde1a,
      rtol=1e-12,atol=1e-12,maxord=5);
  nrow(out)
  ncol(out)
```

The numerical solution to the ODEs is returned in matrix out. In this case, out has the dimensions $nout \times (nx + 1) = 6 \times 9 + 1 = 10$, which are confirmed by the output from nrow(out),ncol(out) (included in the numerical output considered subsequently).

The offset $nx + 1$ is required since the first element of each column has the output t (also in tout), and the $2, ..., nx + 1 = 2, ..., 10$ column elements have the 9 ODE solutions.

The relative and absolute error tolerances are 1e-12. The default values for lsode are 1e-06. Varying this error tolerance gives an indication of the convergence of the solution in t.

- An array u is defined for the numerical $u(x, t)$.

```
#
# Allocate array for u(x,t)
  nx=nx+2;
  u=matrix(0,nt,nx);
```

Note that again nx=11.

---

[7]lsode uses the full ODE Jacobian matrix which for the present case is $9 \times 9 = 81$. This small Jacobian matrix does not place a limitation on the numerical integration by lsode. For larger ODE systems, the sparse integrator lsodes can be used, as demonstrated with examples later in the book.

- The solutions of the 9 ODEs returned in out by lsode are placed in u.

```
#
# u(x,t), x ne xl,xu
  for(i in 1:nt){
    for(j in 2:(nx-1)){
      u[i,j]=out[i,j];
    }
  }
```

Note that j=2,3,...,10 corresponding to the solution of the 9 ODEs.

- BCs (1.3c,d) are used to set $u(x = 0, t), u(x = 1, t)$.

```
#
# Reset boundary values
  for(i in 1:nt){
   u[i,1]=g_0(tout[i]);
   u[i,nx]=g_L(tout[i]);
   }
```

u now has 6 values of $u(x, t)$ for t and 11 values for x for both the numerical and graphical output (a total of $(6)(11) = 66$ values of $u(x, t)$).

- The analytical $u(x, t)$ is used to compute the error in the numerical solution.

```
#
# Numerical, analytical solutions, maximum difference
  uap=matrix(0,nt,nx);
  for(i in 1:nt){
    for(j in 1:nx){
      uap[i,j]=ua((j-1)*dx,(i-1)*dt);
    }
  max_err=max(abs(u-uap));
  }
```

The absolute maximum error is determined by the abs and max utilities.

- The numerical and analytical solutions, and the error are displayed.

```
#
# Tabular numerical, analytical solutions,
```

```
# difference
  cat(sprintf("\n       t      x     u(x,t)
                   ua(x,t)        diff"));
  for(i in 1:nt){
  iv=seq(from=1,to=nx,by=1);
  for(j in iv){
    cat(sprintf("\n %6.2f%6.2f%10.5f%10.5f%12.3e",
      tout[i],xj[j],u[i,j],uap[i,j],u[i,j]-uap[i,j]));
  }
  cat(sprintf("\n"));
  }
```

iv=seq(from=1,to=nx,by=1) is used to vary the number of output lines in x, in this case, every line with by=1.

- The numerical solution is plotted with lines (matplot) and the analytical solution is superimposed as points (matpoints).

```
#
# Plot numerical, analytical solutions
  matplot(xj,t(u),type="l",lwd=2,col="black",lty=1,
    xlab="x",ylab="u(x,t)",main="");
  matpoints(xj,t(uap),pch="o",col="black");
```

t(u),t(uap) (transposes) are required so that the number of rows of u, uap equals the number of rows (elements) of xj. For nt=6, nx=11, u,uap are 6×11 (6 rows $\times$ 11 columns) and a transpose to 11×6 is required.

- The maximum absolute error along the solution is displayed.

```
#
# Display maximum error
  cat(sprintf(" Maximum error = %6.2e \n",max_err));
```

- The absolute error at $t = t_f$ is plotted as a function of x.

```
#
# Plot error at t = tf
  err_1=abs(u[nt,]-ua(xj[1:nx],tf));
  plot(xj,err_1,type="l",xlab="x",
      ylab="Max Error at t = tf",
      main="",col="black")
```

- The numbers of calls to pde1a is displayed as a measure of the computational effort required to compute the solution.

```
#
# Calls to ODE routine
   cat(sprintf("\n\n  ncall = %3d\n",ncall));
```

The ODE/MOL routine pde1a called by lsode is discussed next.

1.2.2   SUBORDINATE ODE/MOL ROUTINE

Listing 1.2: ODE/MOL routine for eqs. (1.3a)–(1.3g)

```
  pde1a=function(t,u,parms){
#
# Function pde1a computes the derivative
# vector of the ODEs approximating the
# PDE
#
# Allocate the vector of the ODE
# derivatives
  ut=rep(0,nx);
#
# Boundary approximations of uxx
  uxx=NULL;
  uxx_0=2*g_0(t)-5*u[1]+4*u[2]-u[3];
  uxx[1]=u[2]-2*u[1]+g_0(t);
  uxx[nx]=g_L(t)-2*u[nx]+u[nx-1];
#
# Interior approximation of uxx
  for(k in 2:(nx-1)){
    uxx[k]=u[k+1]-2*u[k]+u[k-1];
  }
#
# PDE
#
# Step through ODEs
  for(j in 1:nx){
#
#   First term in series approximation of
#   fractional derivative
```

```
      ut[j]=uxx_0*A[j,1];
#
#    Subsequent terms in series approximation
#    of fractional derivative
     for(k in 1:j){
        ut[j]=ut[j]+uxx[k]*A[j,k+1];
#
#    Next k (next term in series)
     }
#
#    ODE with fractional derivative
     ut[j]=cd*d(xj[j+1],t)*ut[j]+p(xj[j+1],t);
#
# Next j (next ODE)
   }
#
# Increment calls to pde1a
   ncall <<- ncall+1;
#
# Return derivative vector of ODEs
   return(list(c(ut)));
   }
```

We can note the following details about pde1a.

- The function is defined.

```
   pde1a=function(t,u,parms){
#
# Function pde1a computes the derivative
# vector of the ODEs approximating the
# PDE
```

t is the current value of t in eq. (1.3a). u is the 9-vector of ODE/MOL dependent variables. parm is an argument to pass parameters to pde1a (unused, but required in the argument list). The arguments must be listed in the order stated to properly interface with lsode called in the main program. The derivative vector of the LHS of eq. (1.3a) is calculated next and returned to lsode.

- An array ut is defined (allocated) for the 9-vector of derivatives of u as defined by eq. (1.3a) (nx=9 as defined in the main program of Listing 1.1 prior to the call to lsode).

```
#
# Allocate the vector of the ODE
# derivatives
  ut=rep(0,nx);
```

- The integer derivative $\dfrac{\partial^2 u}{\partial x^2}$ = uxx for use in eq. (1.1) ($n = 2$) is computed.

```
#
# Boundary approximations of uxx
  uxx=NULL;
  uxx_0=2*g_0(t)-5*u[1]+4*u[2]-u[3];
  uxx[1]=u[2]-2*u[1]+g_0(t);
  uxx[nx]=g_L(t)-2*u[nx]+u[nx-1];
```

This code requires some additional explanation.

  – uxx is nulled and then elements of this vector are added.

```
    uxx=NULL;
```

  – $-\dfrac{\partial^2 u(x = 0, t)}{\partial x^2}$ is computed with the four point FD of eq. (1.2i).

```
    uxx_0=2*g_0(t)-5*u[1]+4*u[2]-u[3];
```

  The factor $\dfrac{1}{\Delta x^2}$ of eq. (1.2i) is included in the coefficient cd.

```
  cd=dx^(-alpha)/gamma(4-alpha)
```

  defined in the main program of Listing 1.1, i.e., from eq. (1.2j),

  $$\frac{\Delta x^{2-\alpha}}{\Delta x^2} = \Delta x^{-\alpha}$$

  – $-\dfrac{\partial^2 u(x = dx, t)}{\partial x^2}$ is computed with the three point FD of eq. (1.2h).

```
    uxx[1]=u[2]-2*u[1]+g_0(t);
```

  BC (1.3c) is used at $x = 0$. Again, $\dfrac{1}{\Delta x^2}$ is included in the coefficient cd.

  – $-\dfrac{\partial^2 u(x = 1 - dx, t)}{\partial x^2}$ is computed with the three point FD of eq. (1.2h).

```
    uxx[nx]=g_L(t)-2*u[nx]+u[nx-1];
```

  BC (1.3c) is used at $x = 1$.

- $\dfrac{\partial^2 u(x,t)}{\partial x^2}$ is computed at $x = x_2, ..., x_{nx-1}$

```
#
# Interior approximation of uxx
  for(k in 2:(nx-1)){
    uxx[k]=u[k+1]-2*u[k]+u[k-1];
  }
```

This completes the calculation of $\dfrac{\partial^2 u(x,t)}{\partial x^2}$ over the interval $0 \leq x \leq 1$.

- The first term in the series approximation of the fractional derivative of eq. (1.3a) is computed according to eq. (1.2j).

```
#
# PDE
#
# Step through ODEs
  for(j in 1:nx){
#
#   First term in series approximation of
#   fractional derivative
    ut[j]=uxx_0*A[j,1];
```

- Subsequent terms are computed and added to the running sum of eq. (1.2j).

```
#
#   Subsequent terms in series approximation
#   of fractional derivative
    for(k in 1:j){
      ut[j]=ut[j]+uxx[k]*A[j,k+1];
#
#   Next k (next term in series)
    }
```

This completes the approximation of the fractional derivative in eq. (1.3a) (in vector ut[j]).

- The fractional derivative is used in the ODE/MOL approximation of eq. (1.3a) at point j in x.

```
#
#   ODE with fractional derivative
    ut[j]=cd*d(xj[j+1],t)*ut[j]+p(xj[j+1],t);
#
# Next j (next ODE)
    }
```

- The counter for the calls to pde1a is incremented and returned to the main program of Listing 1.1 by <<-.

```
#
# Increment calls to pde1a
  ncall <<- ncall+1;
```

- The vector of ODEs derivatives in t is returned to lsode as a list (required by lsode for the next step along the solution in t.

```
#
# Return derivative vector of ODEs
  return(list(c(ut)));
  }
```

The main program of Listing 1.1 and the ODE/MOL routine pde1a of Listing 1.2 can now be used to compute a numerial solution to eqs. (1.3a)–(1.3g).

1.2.3   MODEL OUTPUT

Abbreviated numerical output from the R routines in Listings 1.1, 1.2 is shown in Table 1.1.
   We can note the following details about the output in Table 1.1.

Table 1.1: Abbreviated output for eqs. (1.3a)–(1.3g) (*Continues.*)

[1] 6

[1] 10

t	x	u(x,t)	ua(x,t)	diff
0.00	0.00	0.84147	0.84147	0.000e+00
0.00	0.10	0.84989	0.84989	0.000e+00
0.00	0.20	0.87513	0.87513	0.000e+00
0.00	0.30	0.91720	0.91720	0.000e+00
0.00	0.40	0.97611	0.97611	0.000e+00
0.00	0.50	1.05184	1.05184	0.000e+00
0.00	0.60	1.14440	1.14440	0.000e+00
0.00	0.70	1.25379	1.25379	0.000e+00
0.00	0.80	1.38001	1.38001	0.000e+00
0.00	0.90	1.52306	1.52306	0.000e+00
0.00	1.00	1.68294	1.68294	0.000e+00
0.20	0.00	0.93204	0.93204	0.000e+00
0.20	0.10	0.94136	0.94136	-1.863e-12
0.20	0.20	0.96932	0.96932	-1.573e-12
0.20	0.30	1.01592	1.01592	-1.442e-12
0.20	0.40	1.08117	1.08117	-1.495e-12
0.20	0.50	1.16505	1.16505	-1.850e-12
0.20	0.60	1.26757	1.26757	-2.595e-12
0.20	0.70	1.38874	1.38874	-3.539e-12
0.20	0.80	1.52854	1.52854	-4.130e-12
0.20	0.90	1.68699	1.68699	-2.964e-12
0.20	1.00	1.86408	1.86408	0.000e+00

```
             .                    .
             .                    .
             .                    .

    Output for t = 0.4,0.6,0.8 removed
             .                    .
             .                    .
             .                    .
```

Table 1.1: (*Continued.*) Abbreviated output for eqs. (1.3a)–(1.3g)

```
1.00  0.00   0.90930    0.90930    0.000e+00
1.00  0.10   0.91839    0.91839   -4.696e-12
1.00  0.20   0.94567    0.94567   -6.040e-12
1.00  0.30   0.99113    0.99113   -7.208e-12
1.00  0.40   1.05479    1.05479   -7.644e-12
1.00  0.50   1.13662    1.13662   -7.658e-12
1.00  0.60   1.23664    1.23664   -7.173e-12
1.00  0.70   1.35485    1.35485   -6.206e-12
1.00  0.80   1.49125    1.49125   -4.722e-12
1.00  0.90   1.64583    1.64583   -2.695e-12
1.00  1.00   1.81859    1.81859    0.000e+00

ncall = 226
```

- The solution array out from lsode has the dimensions 6×10 as discussed previously.

   ```
   [1] 6
   ```

   ```
   [1] 10
   ```

   The column dimension is $9 + 1 = 10$ since t is included as the first element in each column, followed by the solutions to the 9 ODEs.

- The ICs for the numerical and analytical solutions at $t = 0$ are the same since the analytical solution of eq. (1.3g) is used to define both ICs.

- 11 values of x correspond to nx=11 at each value of t.

- 6 values of t correspond to nout=6.

- The numerical and analytical BCs (1.3c,d) are identical since the analytical solution of eq. (1.3g) is used to define the BCs, e.g., at $t = 0.20$,

   ```
   0.20  0.00   0.93204   0.93204   0.000e+00
   ```

   ```
   0.20  1.00   1.86408   1.86408   0.000e+00
   ```

- The error in the numerical solution is less than 10^{-11} which is probably better than expected with only 11 points in x.

- The solution does not appear to approach a steady state which is expected since the boundary values at $x = 0, 1$ are changing with t, and this produces continuing changes in the solution with t.

- The computational effort is modest.

      ncall = 226

The graphical output is in Figs. 1.1a,b.

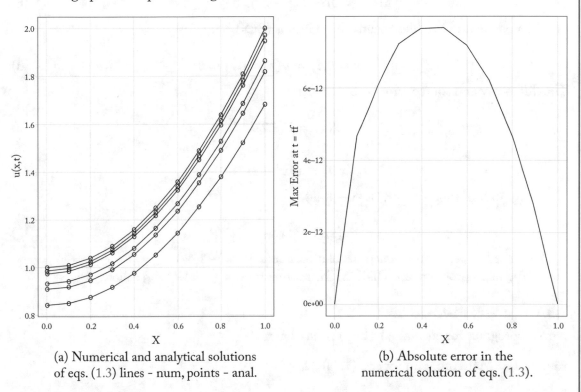

(a) Numerical and analytical solutions
of eqs. (1.3) lines - num, points - anal.

(b) Absolute error in the
numerical solution of eqs. (1.3).

Figure 1.1: (a) Numerical and analytical solutions of eqs. (1.3a)–(1.3g) and (b) absolute error in the numerical solution of eqs. (1.3a)–(1.3g).

Figures 1.1a,b generally reflect the properties of Table 1.1 (e.g., the IC and BCs, the agreement between the numerical and analytical solutions).

In summary, the algorithm of eqs. (1.2a)–(1.2j) has provided an accurate numerical solution to the SFPDE eq. (1.2a) with modest computational effort. We can now consider other examples of SFPDEs. In particular, the main program and ODE/MOL routine of Listings 1.1, 1.2 can be used as templates for new applications. This procedure is demonstrated with the example that follows.

1.3   COMPUTER ROUTINES, EXAMPLE 2

The following SFPDE model is a variant of eqs. (1.3a)–(1.3g) ([1], p1140).

$$\frac{\partial u}{\partial t} = d(x)\frac{\partial^{1.8}u}{\partial x^{1.8}} + p(x,t) \tag{1.4a}$$

with an initial condition (IC)

$$u(x,t=0) = (x^2 - x^3); \ 0 \le x \le 1 \tag{1.4b}$$

and homogeneous Dirichlet boundary conditions (BCs)

$$u(x=0,t) = u(x=1,t) = 0 \tag{1.4c,d}$$

The diffusivity and source term in eq. (1.4a) are

$$d(x) = \Gamma(1.2)x^{1.8} \tag{1.4e}$$

$$p(x,t) = (6x^3 - 3x^2)e^{-t} \tag{1.4f}$$

The analytical solution for eqs. (1.4a)–(1.4g) is

$$u_a(x,t) = (x^2 - x^3)e^{-t} \tag{1.4g}$$

which is used to verify the numerical solution of eqs. (1.4a)–(1.4g).

The main program and ODE/MOL routine for eqs. (1.4a)–(1.4g) follow.

1.3.1   MAIN PROGRAM

A main program for eqs. (1.4a)–(1.4g) is listed next.

Listing 1.3: Main program for eqs. (1.4a)–(1.4g)

```
#
# SFPDE
#
#    ut=d(x)*(d^alpha u/dx^alpha)+p(x,t)
#
#    xl < x < xu, 0 < t < tf, xl=0, xu=1
#
#    u(x,t=0)=(x^2-x^3)
#
#    u(x=xl,t)=u(x=xu,t)=0
#
```

```
#    d(x)=gamma(1.2)*x^1.8
#
#    p(x,t)=(6x^3-3x^2)*e^(-t)
#
#    ua(x,t)=(x^2-x^3)*e^(-t)
#
# Delete previous workspaces
  rm(list=ls(all=TRUE))
#
# Access functions for numerical solution
  library("deSolve");
  setwd("f:/fractional/sfpde/chap1");
  source("pde1b.R");
#
# Parameters
  alpha-1.8;
#
# d(x)
  d=function(x,t) gamma(1.2)*x^(1.8);
#
# p(x,t)
  p=function(x,t) (6*x^3-3*x^2)*exp(-t);
#
# Analytical solution
  ua=function(x,t) (x^2-x^3)*exp(-t);
#
# Initial condition function (IC)
  f=function(x) ua(x,0);

# Boundary condition functions (BCs)
  g_0=function(t) 0;
  g_L=function(t) 0;
#
# Spatial grid
  xl=0;xu=1;nx=21;dx=(xu-xl)/(nx-1);
  xj=seq(from=xl,to=xu,by=dx);
  cd=dx^(-alpha)/gamma(4-alpha);
#
# Independent variable for ODE integration
```

```
  t0=0;tf=1;nt=6;dt=(tf-t0)/(nt-1);
  tout=seq(from=t0,to=tf,by=dt);
#
# a_jk coefficients
  A=matrix(0,nrow=nx-2,ncol=nx-1);
  for(j in 1:(nx-2)){
    for(k in 0:j){
    if (k==0){
      A[j,k+1]=(j-1)^(3-alpha)-j^(2-alpha)*(j-3+alpha);
    } else if (1 <= k && k<=j-1){
      A[j,k+1]=(j-k+1)^(3-alpha)-2*(j-k)^(3-alpha)+(j-k-1)^(3-
          alpha);
    } else
      A[j,k+1]=1;
    }
  }
#
# Initial condition
  nx=nx-2;
  u0=rep(0,nx);
  for(j in 1:nx){
    u0[j]=f(xj[j+1]);}
  ncall=0;
#
# ODE integration
  out=lsode(y=u0,times=tout,func=pde1b,
      rtol=1e-12,atol=1e-12,maxord=5);
  nrow(out)
  ncol(out)
#
# Allocate array for u(x,t)
  nx=nx+2;
  u=matrix(0,nt,nx);
#
# u(x,t), x ne xl,xu
  for(i in 1:nt){
    for(j in 2:(nx-1)){
      u[i,j]=out[i,j];
    }
```

```
  }
#
# Reset boundary values
  for(i in 1:nt){
   u[i,1]=0;
  u[i,nx]=0;
  }
#
# Numerical, analytical solutions, maximum difference
  uap=matrix(0,nt,nx);
  for(i in 1:nt){
    for(j in 1:nx){
      uap[i,j]=ua((j-1)*dx,(i-1)*dt);
    }
  max_err=max(abs(u-uap));
  }
#
# Tabular numerical, analytical solutions,
# difference
  cat(sprintf("\n        t       x      u(x,t)
                    ua(x,t)          diff"));
  for(i in 1:nt){
  iv=seq(from=1,to=nx,by=2);
  for(j in iv){
    cat(sprintf("\n %6.2f%6.2f%10.5f%10.5f%12.3e",
      tout[i],xj[j],u[i,j],uap[i,j],u[i,j]-uap[i,j]));
  }
  cat(sprintf("\n"));
  }
#
# Plot numerical, analytical solutions
  matplot(xj,t(u),type="l",lwd=2,col="black",lty=1,
    xlab="x",ylab="u(x,t)",main="");
  matpoints(xj,t(uap),pch="o",col="black");
#
# Display maximum error
  cat(sprintf(" Maximum error = %6.2e \n",max_err));
#
# Plot error at t = tf
```

```
   err_1=abs(u[nt,]-ua(xj[1:nx],tf));
   plot(xj,err_1,type="l",xlab="x",
       ylab="Max Error at t = tf",
       main="",col="black")
#
# Calls to ODE routine
   cat(sprintf("\n\n  ncall = %3d\n",ncall));
```

The main programs of Listings 1.1 and 1.3 are similar, so only the differences are considered.

- The ODE/MOl routine is pde1b.

```
   #
   # SFPDE
   #
   #   ut=d(x)*(d^alpha u/dx^alpha)+p(x,t)
   #
   #   xl < x < xu, 0 < t < tf, xl=0, xu=1
   #
   #   u(x,t=0)=(x^2-x^3)
   #
   #   u(x=xl,t)=u(x=xu,t)=0
   #
   #   d(x)=gamma(1.2)*x^1.8
   #
   #   p(x,t)=(6x^3-3x^2)*e^(-t)
   #
   #   ua(x,t)=(x^2-x^3)*e^(-t)
   #
   # Delete previous workspaces
     rm(list=ls(all=TRUE))
   #
   # Access functions for numerical solution
     library("deSolve");
     setwd("f:/fractional/sfpde/chap1");
     source("pde1b.R");
```

- Equation (1.4a) is a SFPDE of order 1.8.

```
   #
   # Parameters
     alpha=1.8;
```

- The elements of eqs. (1.4a)–(1.4g) are programmed, including analytical eq. (1.4g).

```
#
# d(x)
  d=function(x,t) gamma(1.2)*x^(1.8);
#
# p(x,t)
  p=function(x,t) (6*x^3-3*x^2)*exp(-t);
#
# Analytical solution
  ua=function(x,t) (x^2-x^3)*exp(-t);
#
# Initial condition function (IC)
  f=function(x) ua(x,0);

# Boundary condition functions (BCs)
  g_0=function(t) 0;
  g_L=function(t) 0;
```

- 21 points are used in the spatial grid (to improve the spatial resolution of the solution).

```
#
# Spatial grid
  xl=0;xu=1;nx=21;dx=(xu-xl)/(nx-1);
  xj=seq(from=xl,to=xu,by=dx);
  cd=dx^(-alpha)/gamma(4-alpha);
```

- pde1b is called by lsode.

```
#
# ODE integration
  out=lsode(y=u0,times=tout,func=pde1b,
      rtol=1e-12,atol=1e-12,maxord=5);
  nrow(out)
  ncol(out)
```

- The solution at the boundaries $x = 0, 1$ is reset as homogeneous Dirichlet BCs (the boundary values are not returned by lsode which returns only solutions to ODEs).

```
#
# Reset boundary values
```

```
      for(i in 1:nt){
       u[i,1]=0;
      u[i,nx]=0;
      }
```

- The solution for every second value of x is displayed.

```
      iv=seq(from=1,to=nx,by=2);
```

Otherwise, the main program of Listing 1.1 is unchanged.

1.3.2   SUBORDINATE ODE/MOL ROUTINE

pde1b is listed next.

Listing 1.4: ODE/MOL routine for eqs. (1.4a)–(1.4g)

```
  pde1b=function(t,u,parms){
#
# Function pde1b computes the derivative
# vector of the ODEs approximating the
# PDE
#
# Allocate the vector of the ODE
# derivatives
  ut=rep(0,nx);
#
# Boundary approximations of uxx
  uxx=NULL;
  uxx_0=2*g_0(t)-5*u[1]+4*u[2]-u[3];
  uxx[1]=u[2]-2*u[1]+g_0(t);
  uxx[nx]=g_L(t)-2*u[nx]+u[nx-1];
#
# Interior approximation of uxx
  for(k in 2:(nx-1)){
    uxx[k]=u[k+1]-2*u[k]+u[k-1];
  }
#
# PDE
#
# Step through ODEs
  for(j in 1:nx){
```

```
#
#    First term in series approximation of
#    fractional derivative
     ut[j]=uxx_0*A[j,1];
#
#    Subsequent terms in series approximation
#    of fractional derivative
     for(k in 1:j){
       ut[j]=ut[j]+uxx[k]*A[j,k+1];
#
#    Next k (next term in series)
     }
#
#    ODE with fractional derivative
     ut[j]=cd*d(xj[j+1],t)*ut[j]+p(xj[j+1],t);
#
# Next j (next ODE)
   }
#
# Increment calls to pde1b
  ncall <<- ncall+1;
#
# Return derivative vector of ODEs
  return(list(c(ut)));
  }
```

pde1b is the same as pde1a of Listing 1.2 except for the name.

```
  pde1b=function(t,u,parms){
#
# Function pde1b computes the derivative
# vector of the ODEs approximating the
# PDE
```

That is, the coding of pde1a, pde1b is generic for a single SFPDE.

1.3.3   MODEL OUTPUT

Abbreviated numerical output from the R routines in Listings 1.3, 1.4 is shown in Table 1.2.
We can note the following details about the output shown in Table 1.2.

Table 1.2: Abbreviated output for for eqs. (1.4a)–(1.4g) (*Continues.*)

```
[1] 6

[1] 20
```

t	x	u(x,t)	ua(x,t)	diff
0.00	0.00	0.00000	0.00000	0.000e+00
0.00	0.10	0.00900	0.00900	0.000e+00
0.00	0.20	0.03200	0.03200	0.000e+00
0.00	0.30	0.06300	0.06300	0.000e+00
0.00	0.40	0.09600	0.09600	0.000e+00
0.00	0.50	0.12500	0.12500	0.000e+00
0.00	0.60	0.14400	0.14400	0.000e+00
0.00	0.70	0.14700	0.14700	0.000e+00
0.00	0.80	0.12800	0.12800	0.000e+00
0.00	0.90	0.08100	0.08100	0.000e+00
0.00	1.00	0.00000	0.00000	0.000e+00
0.20	0.00	0.00000	0.00000	0.000e+00
0.20	0.10	0.00737	0.00737	6.124e-13
0.20	0.20	0.02620	0.02620	1.982e-12
0.20	0.30	0.05158	0.05158	3.442e-12
0.20	0.40	0.07860	0.07860	4.500e-12
0.20	0.50	0.10234	0.10234	4.944e-12
0.20	0.60	0.11790	0.11790	4.779e-12
0.20	0.70	0.12035	0.12035	4.129e-12
0.20	0.80	0.10480	0.10480	3.080e-12
0.20	0.90	0.06632	0.06632	1.634e-12
0.20	1.00	0.00000	0.00000	0.000e+00

```
            .                        .
            .                        .
            .                        .

      Output for t = 0.4,0.6,0.8 removed
            .                        .
            .                        .
            .                        .
```

Table 1.2: (*Continued.*) Abbreviated output for for eqs. (1.4a)–(1.4g)

```
1.00  0.00   0.00000   0.00000   0.000e+00
1.00  0.10   0.00331   0.00331   4.476e-12
1.00  0.20   0.01177   0.01177   9.683e-12
1.00  0.30   0.02318   0.02318   1.295e-11
1.00  0.40   0.03532   0.03532   1.423e-11
1.00  0.50   0.04598   0.04598   1.387e-11
1.00  0.60   0.05297   0.05297   1.224e-11
1.00  0.70   0.05408   0.05408   9.706e-12
1.00  0.80   0.04709   0.04709   6.618e-12
1.00  0.90   0.02980   0.02980   3.286e-12
1.00  1.00   0.00000   0.00000   0.000e+00

ncall = 311
```

- The solution array out from lsode has the dimensions 6×20 (for 19 ODEs).

  [1] 6

  [1] 20

  The column dimension is $19 + 1 = 20$ since t is included as the first element in each column, followed by the solutions to the 19 ODEs.

- The ICs for the numerical and analytical solutions at $t = 0$ are the same since the analytical solution of eq. (1.4g) is used to define both ICs.

- 21 values of x correspond to nx=21 at each value of t, but every second value of x is displayed as explained previously.

- 6 values of t correspond to nout=6.

- The numerical and analytical BCs (1.4c,d) are identical (zero or homogeneous) since the analytical solution of eq. (1.4g) is also homogeneous at $x = 0, 1$.

- The error in the numerical solution is less than 10^{-10} which is probably better than expected with only 21 points in x.

- The solution appears to approach a zero steady state $u(x, t \to \infty) = 0$ which is expected from the analytical solution of eq. (1.4g). An important detail of the numerical solution is that it remains stable with increasing t (lsode maintains stability).

- The computational effort is modest.

```
ncall = 331
```

The graphical output in Figs. 1.2a,b reflects the solution in Table 1.2 (e.g., the IC and BCs, the agreement between the numerical and analytical solutions).

Figures 1.2a,b generally reflect the properties of Table 1.2 (e.g., the IC and BCs, the agreement between the numerical and analytical solutions).

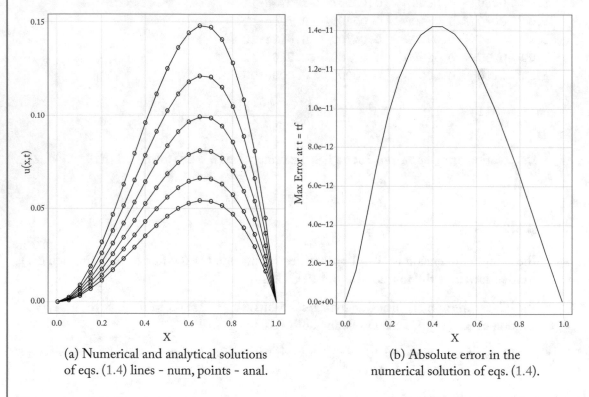

(a) Numerical and analytical solutions
of eqs. (1.4) lines - num, points - anal.

(b) Absolute error in the
numerical solution of eqs. (1.4).

Figure 1.2: (a) Numerical and analytical solutions of eqs. (1.4a)–(1.4g) and (b) Absolute error in the numerical solution of eqs. (1.4a)–(1.4g).

1.3.4   SUMMARY AND CONCLUSIONS

The preceding examples demonstrate the MOL solution of a fractional PDE (eqs. (1.3a), (1.4a)). The routines require only minor modifications for a particular SFPDE. In the next chapter we consider an example SFPDE that demonstrates the effect of varying the order of the fractional derivative.

REFERENCES

[1] Saadatmandi, A., and M. Dehghan (2011), A tau approach for solution of the space fractional diffusion equation, *Computers and Mathematics with Applications*, **62**, 1135–1142. 4, 22

[2] Sousa, E. (2011), Numerical approximations for fractional diffusion equations via splines, *Computers and Mathematics with Applications*, **62**, 938–944. 1, 2, 3

CHAPTER 2

Variation in the Order of the Fractional Derivatives

2.1 INTRODUCTION

The discussion of the two example space factional partial differential differential equations (SF-PDEs) in Chapter 1 is now extended to other applications with the intent of demonstrating some numerical properies of SFPDEs.

2.2 COMPUTER ROUTINES, EXAMPLE 1

The numerical integration of the SFPDE

$$\frac{\partial u}{\partial t} = d(x)\frac{\partial^\alpha u}{\partial x^\alpha} + p(x,t) \tag{2.1a}$$

is discussed in Chapter 1 in terms of two examples, eqs. (1.3a) and (1.4a). The numerical method of lines (MOL) algorithm is validated by comparing the numerical solutions with analytical solutions, eqs. (1.3g) and (1.4g). However, these examples are based on single values of the order of the fractional derivative, $\alpha = 1.5$ for eq. (1.3a), and $\alpha = 1.8$ for eq. (1.4a). We now consider two examples for which α is varied to demonstrate the effect of the order of the fractional derivative.

The particular functions, initial conditions (ICs) and boundary conditions (BCs) for eq. (2.1a) are [1]

$$d(x) = \frac{24}{\Gamma(5+\alpha)}x^\alpha \tag{2.1b}$$

$$p(x,t) = -2e^{-t}x^{(4+\alpha)} \tag{2.1c}$$

The analytical (exact) solution, which is used to verify the numerical solution, is

$$u_a(x,t) = e^{-t}x^{(4+\alpha)} \tag{2.2}$$

From eq. (2.2), the IC is

$$u(x,t=0) = x^{(4+\alpha)} \tag{2.3a}$$

and the Dirchlet BCs are

$$u(x = 0, t) = 0; \; u(x = 1, t) = e^{-t} \qquad\qquad (2.3b,c)$$

The R routines for (2.1), (2.2), and (2.3) follow.

2.2.1   MAIN PROGRAM

Listing 2.1: Main program for eqs. (2.1), (2.2) and (2.3)

```
#
#  SFPDE
#
#    ut=d(x)*(d^alpha u/dx^alpha)+p(x,t)
#
#    xl < x < xu, 0 < t < tf, xl=0, xu=1

#    u(x,t=0)=x^(4+alpha)
#
#    u(x=xl,t)=0; u(x=xu,t)=exp(-t)
#
#    d(x)=24/gamma(5+alpha)*x^alpha
#
#    p(x,t)=-2*exp(-t)*x^(4+alpha)
#
#    ua(x,t)=exp(-t)*x^(4+alpha)
#
# Delete previous workspaces
  rm(list=ls(all=TRUE))
#
# Access functions for numerical solution
  library("deSolve");
  setwd("f:/fractional/sfpde/chap2");
  source("pde1a.R");
#
# Parameters
  for(ncase in 1:5){
    if(ncase==1){alpha=1;}
    if(ncase==2){alpha=1.25;}
    if(ncase==3){alpha=1.5;}
    if(ncase==4){alpha=1.75;}
```

```
      if(ncase==5){alpha=2;}
#
# d(x)
  d=function(x,t) 24/gamma(5+alpha)*x^alpha;
#
# p(x,t)
  p=function(x,t) -2*exp(-t)*x^(4+alpha);
#
# Analytical solution
  ua=function(x,t) exp(-t)*x^(4+alpha);
#
# Initial condition function (IC)
  f=function(x) ua(x,0);
#
# Boundary condition functions (BCs)
  g_0=function(t) ua(xl,t);
  g_L=function(t) ua(xu,t);
#
# Spatial grid
  xl=0;xu=1;nx=41;dx=(xu-xl)/(nx-1);
  xj=seq(from=xl,to=xu,by=dx);
  cd=dx^(-alpha)/gamma(4-alpha);
#
# Independent variable for ODE integration
  t0=0;tf=1;nt=6;dt=(tf-t0)/(nt-1);
  tout=seq(from=t0,to=tf,by=dt);
  ncall=0;
#
# a_jk coefficients
  A=matrix(0,nrow=nx-2,ncol=nx-1);
  for(j in 1:(nx-2)){
    for(k in 0:j){
    if (k==0){
      A[j,k+1]=(j-1)^(3-alpha)-j^(2-alpha)*(j-3+alpha);
    } else if (1 <= k && k<=j-1){
      A[j,k+1]=(j-k+1)^(3-alpha)-2*(j-k)^(3-alpha)+(j-k-1)^(3-
          alpha);
    } else
      A[j,k+1]=1;
```

```
      }
    }
#
# Initial condition
  nx=nx-2;
  u0=rep(0,nx);
  for(j in 1:nx){
    u0[j]=f(xj[j+1]);}
#
# ODE integration
  out=lsode(y=u0,times=tout,func=pde1a,
      rtol=1e-6,atol=1e-6,maxord=5);
  nrow(out)
  ncol(out)
#
# Allocate array for u(x,t)
  nx=nx+2;
  u=matrix(0,nt,nx);
#
# u(x,t), x ne xl,xu
  for(i in 1:nt){
    for(j in 2:(nx-1)){
      u[i,j]=out[i,j];
    }
  }
#
# Reset boundary values
  for(i in 1:nt){
   u[i,1]=g_0(tout[i]);
  u[i,nx]=g_L(tout[i]);
  }
#
# Numerical, analytical solutions, maximum difference
  uap=matrix(0,nt,nx);
  for(i in 1:nt){
    for(j in 1:nx){
      uap[i,j]=ua((j-1)*dx,(i-1)*dt);
    }
  max_err=max(abs(u-uap));
```

```
  }
#
# Tabular numerical, analytical solutions,
# difference
  cat(sprintf("\n\n   alpha = %4.2f\n",alpha));
  cat(sprintf("\n      t      x      u(x,t)    ua(x,t)        diff"))
      ;
  for(i in 1:nt){
  iv=seq(from=1,to=nx,by=4);
  for(j in iv){
    cat(sprintf("\n %6.2f%6.2f%10.5f%10.5f%12.3e",
      tout[i],xj[j],u[i,j],uap[i,j],u[i,j]-uap[i,j]));
  }
  cat(sprintf("\n"));
  }
#
# Plot numerical, analytical solutions
  matplot(xj,t(u),type="l",lwd=2,col="black",lty=1,
    xlab="x",ylab="u(x,t)",main="");
  matpoints(xj,t(uap),pch="o",col="black");
#
# Display maximum error
  cat(sprintf("\n   Maximum error = %6.2e \n",max_err));
#
# Plot error at t = tf
  err_1=abs(u[nt,]-ua(xj[1:nx],tf));
  plot(xj,err_1,type="l",xlab="x",
      ylab="Max Error at t = tf",
      main="",col="black")
#
# Calls to ODE routine
  cat(sprintf("\n   ncall = %3d\n",ncall));
#
# Next alpha (ncase)
  }
```

The main program of Listing 2.1 is similar to that of Listing 1.1, so only the differences are considered here.

    • The ODE/MOL routine is again pde1a.R.

```
#
# Access functions for numerical solution
  library("deSolve");
  setwd("f:/fractional/sfpde/chap2");
  source("pde1a.R");
```

- Five cases are programmed for variations in the fractional derivative of eq. (2.1a), α.

```
#
# Parameters
  for(ncase in 1:5){
    if(ncase==1){alpha=1;}
    if(ncase==2){alpha=1.25;}
    if(ncase==3){alpha=1.5;}
    if(ncase==4){alpha=1.75;}
    if(ncase==5){alpha=2;}
```

The solutions for these five cases are discussed subsequently.

- The diffusivity function $d(x)$ of eq. (2.1b) is programmed.

```
#
# d(x)
  d=function(x,t) 24/gamma(5+alpha)*x^alpha;
```

- The source term $p(x,t)$ of eq. (2.1c) is programmed.

```
#
# p(x,t)
  p=function(x,t) -2*exp(-t)*x^(4+alpha);
```

- IC (2.3a) is programmed.

```
#
# Initial condition function (IC)
  f=function(x) ua(x,0);
```

- BCs (2.3b,c) are programmed.

```
# Boundary condition functions (BCs)
  g_0=function(t) ua(xl,t);
  g_L=function(t) ua(xu,t);
```

- A spatial grid in x of 41 points is defined for $0 \leq x \leq 1$ so dx=0.025 and xj=0,0.025,...,1.

```
#
# Spatial grid
  xl=0;xu=1;nx=41;dx=(xu-xl)/(nx-1);
  xj=seq(from=xl,to=xu,by=dx);
  cd=dx^(-alpha)/gamma(4-alpha);
```

cd is a coefficient for the diffusion term of eq. (2.1a) used in the PDE programming in pde1a.

- The 41 ODEs are programmed in pde1a called by lsode.

```
#
# ODE integration
  out=lsode(y=u0,times=tout,func=pde1a,
      rtol=1e-6,atol=1e-6,maxord=5);
  nrow(out)
  ncol(out)
```

- The numerical solution in u, and the analytical solution in uap are plotted as lines and points, respectively.

```
#
# Plot numerical, analytical solutions
  matplot(xj,t(u),type="l",lwd=2,col="black",lty=1,
    xlab="x",ylab="u(x,t)",main="");
  matpoints(xj,t(uap),pch="o",col="black");
```

The transposes t( ) are required so that number of rows of u,uap equals the size (length) of xj. The graphical (plotted) output is discussed subsequently.

- The maximum absolute error in the numerical solution is displayed and plotted at $t = t_f = 1$ as a function of x.

```
#
# Display maximum error
  cat(sprintf("\n   Maximum error = %6.2e \n",max_err));
#
# Plot error at t = tf
  err_1=abs(u[nt,]-ua(xj[1:nx],tf));
```

```
        plot(xj,err_1,type="l",xlab="x",
            ylab="Max Error at t = tf",
            main="",col="black")
```

- The number of calls to pde1a is displayed and the current pass of the for in α is completed.

```
#
# Calls to ODE routine
  cat(sprintf("\n   ncall = %3d\n",ncall));
#
# Next alpha (ncase)
  }
```

2.2.2   SUBORDINATE ODE/MOL ROUTINE

The ODE/MOl routine, pde1a, call by lsode in the main program of Listing 2.1 follows.

Listing 2.2: ODE/MOL routine pde1a for eqs. (2.1), (2.2), and (2.3)

```
  pde1a=function(t,u,parms){
#
# Function pde1a computes the derivative
# vector of the ODEs approximating the
# PDE
#
# Allocate the vector of the ODE
# derivatives
  ut=rep(0,nx);
#
# Boundary approximations of uxx
  uxx=NULL;
  uxx_0=2*g_0(t)-5*u[1]+4*u[2]-u[3];
  uxx[1]=u[2]-2*u[1]+g_0(t);
  uxx[nx]=g_L(t)-2*u[nx]+u[nx-1];
#
# Interior approximation of uxx
  for(k in 2:(nx-1)){
    uxx[k]=u[k+1]-2*u[k]+u[k-1];
  }
#
# PDE
#
```

```
# Step through ODEs
  for(j in 1:nx){
#
#    First term in series approximation of
#    fractional derivative
     ut[j]=uxx_0*A[j,1];
#
#    Subsequent terms in series approximation
#    of fractional derivative
     for(k in 1:j){
       ut[j]=ut[j]+uxx[k]*A[j,k+1];
#
#    Next k (next term in series)
     }
#
#    ODE with fractional derivative
     ut[j]=cd*d(xj[j+1],t)*ut[j]+p(xj[j+1],t);
#
# Next j (next ODE)
  }
#
# Increment calls to pde1a
  ncall <<- ncall+1;
#
# Return derivative vector of ODEs
  return(list(c(ut)));
  }
```

pde1a is essentially identical to pde1a in Listing 1.2 so it is not discussed here. In particular, eq. (2.1a) is programmed as

```
   ut[j]=cd*d(xj[j+1],t)*ut[j]+p(xj[j+1],t);
```

cd is defined in Listing 2.1.

2.2.3 MODEL OUTPUT

Abbreviated numerical output from the R routines of Listings 2.1 and 2.2 is shown in Table 2.1.
    We can observe the following details about the output shown in Table 2.1.

- The output is for $\alpha = 1, 1.25, 1.5, 1.75, 2$ as programmed in Listing 2.1.

Table 2.1: Numerical solution to eqs. (2.1), (2.3), and (2.4), $\alpha = 1, 1.25, 1.5, 1.75, 2$ (*Continues.*)

```
alpha = 1.00


      t     x    u(x,t)    ua(x,t)         diff
   0.00  0.00   0.00000    0.00000     0.000e+00
   0.00  0.10   0.00001    0.00001     0.000e+00
   0.00  0.20   0.00032    0.00032     0.000e+00
   0.00  0.30   0.00243    0.00243     0.000e+00
   0.00  0.40   0.01024    0.01024     0.000e+00
   0.00  0.50   0.03125    0.03125     0.000e+00
   0.00  0.60   0.07776    0.07776     0.000e+00
   0.00  0.70   0.16807    0.16807     0.000e+00
   0.00  0.80   0.32768    0.32768     0.000e+00
   0.00  0.90   0.59049    0.59049     0.000e+00
   0.00  1.00   1.00000    1.00000     0.000e+00

                  .                 .
                  .                 .
                  .                 .

       Output for t = 0.2 to 0.8 removed

                  .                 .
                  .                 .
                  .                 .

   1.00  0.00   0.00000    0.00000     0.000e+00
   1.00  0.10   0.00000    0.00000     8.397e-07
   1.00  0.20   0.00013    0.00012     8.492e-06
   1.00  0.30   0.00092    0.00089     2.975e-05
   1.00  0.40   0.00384    0.00377     7.138e-05
   1.00  0.50   0.01164    0.01150     1.401e-04
   1.00  0.60   0.02885    0.02861     2.425e-04
   1.00  0.70   0.06221    0.06183     3.850e-04
   1.00  0.80   0.12109    0.12055     5.424e-04
   1.00  0.90   0.21751    0.21723     2.783e-04
   1.00  1.00   0.36788    0.36788     0.000e+00

Maximum error = 5.48e-04

ncall = 222
```

Table 2.1: (*Continued.*) Numerical solution to eqs. (2.1), (2.3), and (2.4), $\alpha = 1, 1.25, 1.5, 1.75, 2$ (*Continues.*)

```
alpha = 1.25
     .

     .

     .
Output for t = 0 to 1 removed
     .

     .

     .
Maximum error = 3.90e-04

ncall = 222
alpha = 1.50
     .

     .

     .
Output for t = 0 to 1 removed
     .

     .

     .

Maximum error = 2.94e-04

ncall = 262

alpha = 1.75
     .

     .

     .
Output for t = 0 to 1 removed
     .

     .

     .
Maximum error = 2.17e-04

ncall = 222
```

Table 2.1: (*Continued.*) Numerical solution to eqs. (2.1), (2.3), and (2.4), $\alpha = 1, 1.25, 1.5, 1.75, 2$
$\alpha = 1, 1.25, 1.5, 1.75, 2$

```
alpha = 2.00

   t      x     u(x,t)    ua(x,t)         diff
 0.00   0.00   0.00000    0.00000     0.000e+00
 0.00   0.10   0.00000    0.00000     0.000e+00
 0.00   0.20   0.00006    0.00006     0.000e+00
 0.00   0.30   0.00073    0.00073     0.000e+00
 0.00   0.40   0.00410    0.00410     0.000e+00
 0.00   0.50   0.01562    0.01562     0.000e+00
 0.00   0.60   0.04666    0.04666     0.000e+00
 0.00   0.70   0.11765    0.11765     0.000e+00
 0.00   0.80   0.26214    0.26214     0.000e+00
 0.00   0.90   0.53144    0.53144     0.000e+00
 0.00   1.00   1.00000    1.00000     0.000e+00
          .                    .
          .                    .
          .                    .

     Output for t = 0.2 to 0.8 removed
          .                    .
          .                    .
          .                    .

 1.00   0.00   0.00000    0.00000     0.000e+00
 1.00   0.10   0.00000    0.00000     5.045e-08
 1.00   0.20   0.00002    0.00002     8.035e-07
 1.00   0.30   0.00027    0.00027     4.062e-06
 1.00   0.40   0.00152    0.00151     1.280e-05
 1.00   0.50   0.00578    0.00575     3.073e-05
 1.00   0.60   0.01722    0.01716     5.930e-05
 1.00   0.70   0.04337    0.04328     9.134e-05
 1.00   0.80   0.09655    0.09644     1.077e-04
 1.00   0.90   0.19559    0.19551     8.353e-05
 1.00   1.00   0.36788    0.36788     0.000e+00

Maximum error = 1.08e-04

ncall = 223
```

```
#
# Parameters
  for(ncase in 1:5){
    if(ncase==1){alpha=1;}
    if(ncase==2){alpha=1.25;}
    if(ncase==3){alpha=1.5;}
    if(ncase==4){alpha=1.75;}
    if(ncase==5){alpha=2;}
```

- The numerical and analytical solutions are the same for the IC at $t = 0$ since both are defined by the analytical solution of eq. (2.2).

- The output is for $x = 0, 0.025, ..., 1$ as programmed in Listing 2.1.

- The output is for $t = 0, 0.2, ..., 1$ as programmed in Listing 2.1.

- The maximum error of the five cases is for $\alpha = 1$.

```
    Maximum error = 5.48e-04
```

Therefore, all of the errors are bounded to an acceptable level with a spatial grid of 41 points.

- The largest value of `ncall` corresponds to $\alpha = 1.5$

```
    ncall = 262
```

Therefore, the computational effort for the five solutions is modest.

The graphical output is in Figs. 2.1, 2.2, 2.3, 2.4 for $\alpha = 1, 2$ (the graphical output for $\alpha = 1.25, 1.5, 1.75$ is not included to conserve space).

   The plots for the superimposed numerical and analytical solutions, e.g., Figs. 2.1, 2.3, are essentially identical since the BCs constrain the solutions to the same boundary values ($u(0, t) = 0, u(x = 1, t) = e^{-t}$), and the PDE, eq. (2.1a), interpolates between the two boundary values in a way that is essentially independent of α. Therefore, to demonstrate the effect of α in the solution of eq. (2.1a), a second example is now considered.

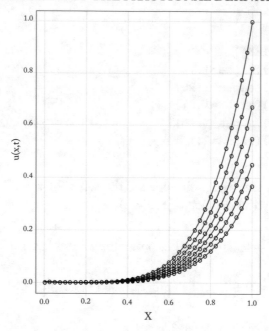

Figure 2.1: Numerical and analytical solutions of eqs. (2.1), (2.2), and (2.3) $\alpha = 1$, lines - num, points - anal.

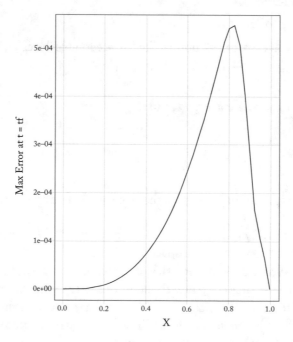

Figure 2.2: Numerical error over $0 \leq x \leq 1$ for $t = t_f = 1$, $\alpha = 1$.

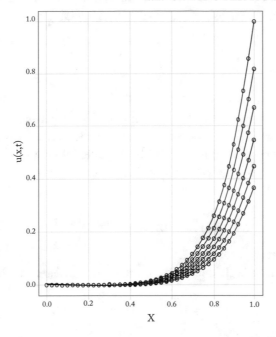

Figure 2.3: Numerical and analytical solutions of eqs. (2.1), (2.2), and (2.3) $\alpha = 2$, lines - num, points - anal.

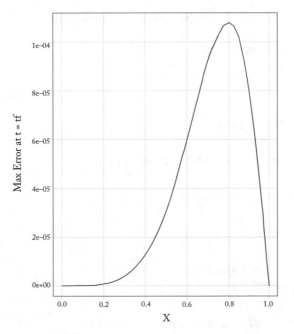

Figure 2.4: Numerical error over $0 \leq x \leq 1$ for $t = t_f = 1$, $\alpha = 2$.

2.3 COMPUTER ROUTINES, EXAMPLE 2

The SFPDE is

$$\frac{\partial u}{\partial t} = \frac{\partial^\alpha u}{\partial x^\alpha} \tag{2.4a}$$

Equation (2.4a) is a fractional form of the integer advection equation for $\alpha = 1$, and the integer diffusion equation for $\alpha = 2$. These special cases are considered subsequently.

An analytical (exact) solution is not readily available for eq. (2.4a) so that only the numerical solution is presented using the same algorithms and coding validated with example 1.

The IC is

$$u(x, t = 0) = e^{-100(x-0.5)^2} \tag{2.4b}$$

a Gaussian function centered at $x = 0.5$.

Homogeneous Dirichlet BCs are selected that are consistent with the IC (2.4b).

$$u(x = 0, t) = u(x = 1, t) = 0 \tag{2.4c,d}$$

The R routines for eqs. (2.4) follow.

2.3.1 MAIN PROGRAM

A main program for eqs. (2.4) follows that is similar to the main program of Listing 2.1.

Listing 2.3: Main program for eqs. (2.4)

```
#
# SFPDE
#
#   ut=d^alpha u/dx^alpha)
#
#   xl < x < xu, 0 < t < tf, xl=0, xu=1
#
#   u(x,t=0)=e^(-c*(x-0.5)^2)
#
#   u(x=xl,t)=0; u(x=xu,t)=0
#
# Delete previous workspaces
  rm(list=ls(all=TRUE))
#
# Access functions for numerical solution
  library("deSolve");
```

```
  setwd("f:/fractional/sfpde/chap2");
  source("pde1b.R");
#
# Parameters
  for(ncase in 1:5){
    if(ncase==1){alpha=1;}
    if(ncase==2){alpha=1.25;}
    if(ncase==3){alpha=1.5;}
    if(ncase==4){alpha=1.75;}
    if(ncase==5){alpha=2;}
#
# Initial condition function (IC)
  f=function(x) exp(-100*(x-0.5)^2);
#
# Boundary condition functions (BCs)
  g_0=function(t) 0;
  g_L=function(t) 0;
#
# Spatial grid
  xl=0;xu=1;nx=51;dx=(xu-xl)/(nx-1);
  xj=seq(from=xl,to=xu,by=dx);
#
# Independent variable for ODE integration
  t0=0;tf=5;nt=6;dt=(tf-t0)/(nt-1);
  tout=seq(from=t0,to=tf,by=dt);
  ncall=0;
#
# a_jk coefficients
  A=matrix(0,nrow=nx-2,ncol=nx-1);
  for(j in 1:(nx-2)){
    for(k in 0:j){
    if (k==0){
      A[j,k+1]=(j-1)^(3-alpha)-j^(2-alpha)*(j-3+alpha);
    } else if (1 <= k && k<=j-1){
      A[j,k+1]=(j-k+1)^(3-alpha)-2*(j-k)^(3-alpha)+(j-k-1)^(3-
         alpha);
    } else
      A[j,k+1]=1;
    }
```

```
    }
#
# Initial condition
  nx=nx-2;
  u0=rep(0,nx);
  for(j in 1:nx){
    u0[j]=f(xj[j+1]);}
#
# ODE integration
  out=lsode(y=u0,times=tout,func=pde1b,
      rtol=1e-6,atol=1e-6,maxord=5);
  nrow(out)
  ncol(out)
#
# Allocate array for u(x,t)
  nx=nx+2;
  u=matrix(0,nt,nx);
#
# u(x,t), x ne xl,xu
  for(i in 1:nt){
    for(j in 2:(nx-1)){
      u[i,j]=out[i,j];
    }
  }
#
# Reset boundary values
  for(i in 1:nt){
   u[i,1]=g_0(tout[i]);
   u[i,nx]=g_L(tout[i]);
   }
#
# Tabular numerical solution
  cat(sprintf("\n\n   alpha = %4.2f\n",alpha));
  cat(sprintf("\n      t      x     u(x,t)"));
  for(i in 1:nt){
  iv=seq(from=1,to=nx,by=5);
  for(j in iv){
    cat(sprintf("\n %6.2f%6.2f%10.5f",
      tout[i],xj[j],u[i,j]));
```

```
  }
  cat(sprintf("\n"));
  }
#
# Plot numerical solution
  matplot(xj,t(u),type="l",lwd=2,col="black",lty=1,
    xlab="x",ylab="u(x,t)",main="");
#
# Calls to ODE routine
  cat(sprintf("\n    ncall = %3d\n",ncall));
#
# Next alpha (ncase)
  }
```

The main program of Listing 2.3 is similar to that of Listing 2.1, so only the differences are considered here.

- The ODE/MOL routine is pde1b.R.

  ```
  #
  # Access functions for numerical solution
    library("deSolve");
    setwd("f:/fractional/sfpde/chap2");
    source("pde1b.R");
  ```

- IC (2.4b) is programmed as

  ```
  #
  # Initial condition function (IC)
    f=function(x) exp(-100*(x-0.5)^2);
  ```

- BCs (2.4c,d) are programmed as

  ```
  #
  # Boundary condition functions (BCs)
    g_0=function(t) 0;
    g_L=function(t) 0;
  ```

- A spatial grid in x of 51 points is progarmmed for $0 \leq x \leq 1$ so xj=0,0.02,...,1.

  ```
  #
  ```

```
# Spatial grid
  xl=0;xu=1;nx=51;dx=(xu-xl)/(nx-1);
  xj=seq(from=xl,to=xu,by=dx);
```

• An interval in t of 6 points is progarmmed for $0 \leq t \leq 5$ so tout=0,1,...,5.

```
#
# Independent variable for ODE integration
  t0=0;tf=5;nt=6;dt=(tf-t0)/(nt-1);
  tout=seq(from=t0,to=tf,by=dt);
  ncall=0;
```

• lsode calls pde1b (considered next).

```
#
# ODE integration
  out=lsode(y=u0,times=tout,func=pde1b,
      rtol=1e-6,atol=1e-6,maxord=5);
  nrow(out)
  ncol(out)
```

The dimensions of out from nrow(out),ncol(out) are considered subsequently.

• The numerical solution is displayed for the five values of α (for every fifth value of x).

```
#
# Tabular numerical solution
  cat(sprintf("\n\n    alpha = %4.2f\n",alpha));
  cat(sprintf("\n       t      x     u(x,t)"));
  for(i in 1:nt){
  iv=seq(from=1,to=nx,by=5);
  for(j in iv){
    cat(sprintf("\n %6.2f%6.2f%10.5f",
      tout[i],xj[j],u[i,j]));
  }
  cat(sprintf("\n"));
  }
```

• The numerical solution is plotted with matplot.

```
#
```

```
# Plot numerical solution
  matplot(xj,t(u),type="l",lwd=2,col="black",lty=1,
    xlab="x",ylab="u(x,t)",main="");
```

The transpose `t(u)` is required so that the number of rows of u is the same as the length of `xj`. u is therefore plotted parametrically in *t* (according to the first dimension of u).

2.3.2   SUBORDINATE ODE/MOL ROUTINE

`pde1b` called by `lsode` is listed next.

Listing 2.4: ODE/MOL routine for eqs. (2.4)

```
  pde1b=function(t,u,parms){
#
# Function pde1b computes the derivative
# vector of the ODEs approximating the
# PDE
#
# Allocate the vector of the ODE
# derivatives
  ut=rep(0,nx);
#
# Boundary approximations of uxx
  uxx=NULL;
  uxx_0=2*g_0(t)-5*u[1]+4*u[2]-u[3];
  uxx[1]=u[2]-2*u[1]+g_0(t);
  uxx[nx]=g_L(t)-2*u[nx]+u[nx-1];
#
# Interior approximation of uxx
  for(k in 2:(nx-1)){
    uxx[k]=u[k+1]-2*u[k]+u[k-1];
  }
#
# PDE
#
# Step through ODEs
  for(j in 1:nx){
#
#    First term in series approximation of
#    fractional derivative
    ut[j]=uxx_0*A[j,1];
```

```
#
#    Subsequent terms in series approximation
#    of fractional derivative
     for(k in 1:j){
        ut[j]=ut[j]+uxx[k]*A[j,k+1];
#
#    Next k (next term in series)
     }
#
# Next j (next ODE)
   }
#
# Increment calls to pde1b
   ncall <<- ncall+1;
#
# Return derivative vector of ODEs
   return(list(c(ut)));
   }
```

pde1b is essentially identical to pde1a in Listing 2.2 so it is not discussed here. In particular, eq. (2.4a) is programmed as

```
        ut[j]=ut[j]+uxx[k]*A[j,k+1];
```

The numerical and graphical output from the R routines of Listings 2.3 and 2.4 follows.

2.3.3   MODEL OUTPUT

Abbreviated numerical output from the R routines of Listings 2.3 and 2.4 is shown in Table 2.2. We can observe the following details about the output shown in Table 2.2.

- The output is for $\alpha = 1, 1.25, 1.5, 1.75, 2$ as programmed in Listing 2.3.

- The output is for $x = 0, 0.02, ..., 1$ as programmed in Listing 2.3.

- The output is for $t = 0, 1, ..., 5$ as programmed in Listing 2.3.

- The numerical solution has the value $u(x = 0.5, t = 0) = 1$ as expected for the Gaussian function of eq. (2.4b).

- BCs (2.4c,d) are consistent with the numerical solution at the interior points in x. That is, the solution near $x = 0, 1$ is close to zero.

- The largest value of ncall corresponds to $\alpha = 1$

Table 2.2: Numerical solution to eqs. (2.4), $\alpha = 1, 1.25, 1.5, 1.75, 2$ (*Continues.*)

```
alpha = 1.00

    t      x      u(x,t)
  0.00   0.00   0.00000
  0.00   0.10   0.00000
  0.00   0.20   0.00012
  0.00   0.30   0.01832
  0.00   0.40   0.36788
  0.00   0.50   1.00000
  0.00   0.60   0.36788
  0.00   0.70   0.01832
  0.00   0.80   0.00012
  0.00   0.90   0.00000
  0.00   1.00   0.00000

            .
            .
            .

Output for t = 1
   to 4 removed

            .
            .
            .

  5.00   0.00    0.00000
  5.00   0.10    0.02636
  5.00   0.20    0.35216
  5.00   0.30    0.98993
  5.00   0.40    0.39407
  5.00   0.50    0.00612
  5.00   0.60   -0.00061
  5.00   0.70   -0.00061
  5.00   0.80   -0.00061
  5.00   0.90   -0.00061
  5.00   1.00    0.00000

ncall = 345
```

Table 2.2: (*Continued.*) Numerical solution to eqs. (2.4), $\alpha = 1, 1.25, 1.5, 1.75, 2$ (*Continues.*)

```
alpha = 1.25
    .

    .

    .

Output for t = 0
  to 5 removed
    .

    .

    .

ncall = 336


alpha = 1.50
    .

    .

    .

Output for t = 0
  to 5 removed
    .

    .

    .

ncall = 330


alpha = 1.75
    .

    .

    .

Output for t = 0
  to 5 removed
    .

    .

    .

ncall = 326
```

Table 2.2: (*Continued.*) Numerical solution to eqs. (2.4), $\alpha = 1, 1.25, 1.5, 1.75, 2$

```
alpha = 2.00

   t     x     u(x,t)
0.00   0.00   0.00000
0.00   0.10   0.00000
0.00   0.20   0.00012
0.00   0.30   0.01832
0.00   0.40   0.36788
0.00   0.50   1.00000
0.00   0.60   0.36788
0.00   0.70   0.01832
0.00   0.80   0.00012
0.00   0.90   0.00000
0.00   1.00   0.00000

            .

            .

            .

Output for t = 1
   to 4 removed

            .

            .

            .

5.00   0.00   0.00000
5.00   0.10   0.00012
5.00   0.20   0.00520
5.00   0.30   0.08052
5.00   0.40   0.42679
5.00   0.50   0.74721
5.00   0.60   0.42679
5.00   0.70   0.08052
5.00   0.80   0.00520
5.00   0.90   0.00012
5.00   1.00   0.00000

ncall = 274
```

```
ncall = 345
```

Therefore, the computational effort for the five solutions is modest.

The graphical output is in Figs. 2.5 to 2.9 for $\alpha = 1, 1.25, 1.5, 1.75, 2$.

The solution near the boundaries at $x = 0, 1$ is close to zero indicating a homogeneous Dirichlet BCs (2.4c,d) are closely approximated. Also, for $\alpha = 1$, eq. (2.4a) is first order in x and is therefore the linear advection equation,

$$\frac{\partial u}{\partial t} = v \frac{\partial u}{\partial x} \tag{2.5a}$$

with IC

$$u(x, t = 0) = f(x) \tag{2.5b}$$

Eq. (2.5a) has the well known solution

$$u(x, t) = f(\lambda); \ \lambda = x + vt \tag{2.5c}$$

that is, the IC translates right to left with velocity v (λ is termed a *Lagrangian* variable). The translation of the Gaussian IC from right to left with increasing t is clear in Fig. 2.5 (the effective velocity v can be estimated from Fig. 2.5 with $\lambda = 0$ corresponding to the maximum of the Gaussian at each t). Also, the maximum has a slightly irregular shape and varying value due to a gridding effect. If nx=51 is increased to nx=101, the gridding effect is effectively eliminated.

Here we observe that the Caputo fractional derivative reduces to the usual first order integer derivative when α is integer one. Eq. (2.5a) is a first order hyperbolic PDE, so that for $\alpha = 1$, the SFPDE (2.4a) is hyperbolic (or convective).

For $\alpha = 1.25, 1.5, 1.75$, eq. (2.4a) becomes less hyperbolic and increasingly parabolic (diffusive), so that the Gaussian function is smoothed with increasing t. That is, eq. (2.4a) is hyperbolic-parabolic.

For $\alpha = 2$, eq. (2.4a) is second order in x and is therefore the linear diffusion equation,

$$\frac{\partial u}{\partial t} = D \frac{\partial^2 u}{\partial x^2} \tag{2.5d}$$

where D is a diffusion coefficient or diffusivity. The center of the IC Gaussian remains stationary at $x = 0$, but is dispersed (smoothed, diffused) in x.

In summary, eq. (2.4a) changes from hyperbolic to parabolic as α changes from 1 to 2. For the intermediate values of $\alpha = 1.25, 1.5, 1.75$, eq. (2.4a) is hyperbolic-parabolic (convective-diffusive).

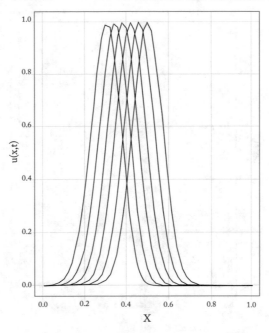

Figure 2.5: Numerical solution of eqs. (2.4), $\alpha = 1$.

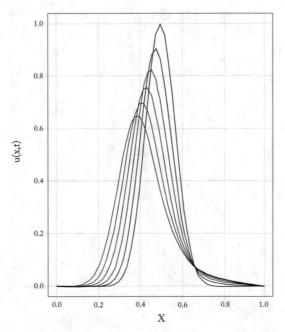

Figure 2.6: Numerical solution of eqs. (2.4), $\alpha = 1.25$.

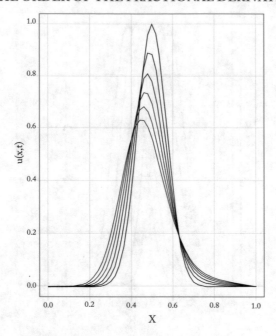

Figure 2.7: Numerical solution of eqs. (2.4), $\alpha = 1.5$.

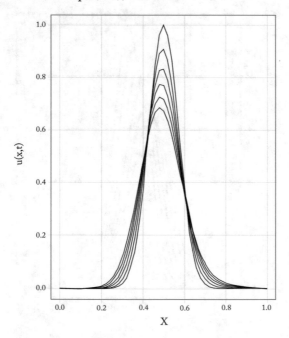

Figure 2.8: Numerical solution of eqs. (2.4), $\alpha = 1.75$.

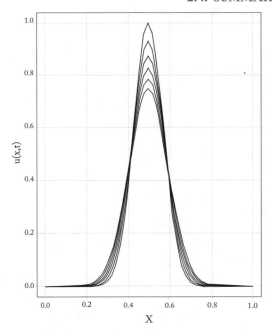

Figure 2.9: Numerical solution of eqs. (2.4), $\alpha = 2$.

2.4 SUMMARY AND DISCUSSION

The preceding examples demonstrate the effect of the order of the fractional derivative

$$\frac{\partial^\alpha u}{\partial x^\alpha}$$

A spectrum of solutions results that might better represent the application (e.g., experimental data) than the corresponding integer order PDEs. Also, other terms, for example, for advection, reaction, coupling between simultaneous SFPDEs, can be added to account for other physical/-chemical phenomena. This variation of the form of the SFPDE is consider subsequently.

The MOL approach to the solution of eqs. (2.4) produced numerical solutions of acceptable accuracy with modest computational effort. Since this approach is numerical, variations in the SFPDEs (e.g., to include additional terms, both linear and nonlinear) will be straightforward as discussed in subsequent chapters.

REFERENCES

[1] Sousa, E. (2011), Numerical approximations for fractional diffusion equations via splines, *Computers and Mathematics with Applications*, **62**, 938–944. 35

CHAPTER 3

Dirichlet, Neumann, Robin BCs

3.1 INTRODUCTION

The previous example applications of space fractional partial differential equations (SFPDEs) are with Dirichlet boundary conditions (BCs), e.g., eqs. (1.3a), (1.3c,d), (1.4a), (1.4c,d) in Chapter 1, eqs. (2.1a)–(2.1c), (2.3b,c) in Chapter 2.

In this chapter, an algorithm is presented for SFPDEs with Dirichlet, Neumann and Robin BCs. This algorithm is tested with SFPDEs with known analytical (exact) solutions.

3.2 EXAMPLE 1, DIRICHLET BCS

The numerical integration of SFPDEs with Dirichlet BCs is illustrated with the following system (eqs. (2.1a)–(2.1c), (2.2), and (2.3a), (2.3b,c) restated here for use with various BCs) [1].

$$\frac{\partial u}{\partial t} = d(x)\frac{\partial^\alpha u}{\partial x^\alpha} + p(x,t) \tag{3.1a}$$

$$d(x) = \frac{24}{\Gamma(5+\alpha)}x^\alpha \tag{3.1b}$$

$$p(x,t) = -2e^{-t}x^{(4+\alpha)} \tag{3.1c}$$

The analytical (exact) solution, which is used to verify the numerical solution, is

$$u_a(x,t) = e^{-t}x^{(4+\alpha)} \tag{3.2}$$

From eq. (3.2), the IC is

$$u(x,t=0) = x^{(4+\alpha)} \tag{3.3a}$$

and the Dirichlet BCs are from eq. (3.2)

$$u(x = x_l, t) = e^{-t}x_l^{(4+\alpha)} = g_0(t)$$

$$u(x = x_u, t) = e^{-t}x_u^{(4+\alpha)} = g_L(t) \tag{3.3b,c}$$

Equations (3.1), (3.2), and (3.3) constitute a test problem that can be used to verify a numerical solution for SFPDEs. To illustrate this verification procedure, the following BCs are used with a special case of the Dirichlet BCs of eqs. (3.3b,c).

$$c_2(t)\frac{\partial u(x = x_l, t)}{\partial x} + c_1(t)u(x = x_l, t) = g_0(t) \tag{3.4a}$$

$$c_4(t)\frac{\partial u(x = x_u, t)}{\partial x} + c_3(t)u(x = x_u, t) = g_L(t) \tag{3.4b}$$

For eqs. (3.3b,c), $c_1(t) = 1$, $c_2(t) = 0$, $c_3(t) = 1$, $c_4(t) = 0$. R routines for eqs. (3.1) to (3.6) follow.

3.2.1 MAIN PROGRAM

Listing 3.1: Main program for eqs. (3.1) to (3.6)

```
#
# SFPDE
#
#    ut=d(x)*(d^alpha u/dx^alpha)+p(x,t)
#
#    xl < x < xu, 0 < t < tf, xl=0, xu=1
#
#    u(x,t=0)=x^(4+alpha)
#
#    u(x=xl,t)=0; u(x=xu,t)=exp(-t)
#
#    d(x)=24/gamma(5+alpha)*x^alpha
#
#    p(x,t)=-2*exp(-t)*x^(4+alpha)
#
#    ua(x,t)=exp(-t)*x^(4+alpha)
#
# Delete previous workspaces
  rm(list=ls(all=TRUE))
#
# Access functions for numerical solution
  library("deSolve");
  setwd("f:/fractional/sfpde/chap3/dirichlet");
  source("pde1a.R");
#
# Parameters
```

```
  alpha=1;
#
# d(x)
  d=function(x) 24/gamma(5+alpha)*x^alpha;
#
# p(x,t)
  p=function(x,t) -2*exp(-t)*x^(4+alpha);
#
# Analytical solution
  ua=function(x,t) exp(-t)*x^(4+alpha);
#
# Initial condition function (IC)
  f=function(x) ua(x,0);
#
# Boundary condition functions (BCs)
  g_0=function(t) ua(xl,t);
  g_L=function(t) ua(xu,t);
#
# Boundary condition coefficients
  c_1=function(t) 1;
  c_2=function(t) 0;
  c_3=function(t) 1;
  c_4=function(t) 0;
#
# Spatial grid
  xl=0;xu=1;nx=41;dx=(xu-xl)/(nx-1);
  xj=seq(from=xl,to=xu,by=dx);
  cd=dx^(-alpha)/gamma(4-alpha);
#
# Independent variable for ODE integration
  t0=0;tf=1;nt=6;dt=(tf-t0)/(nt-1);
  tout=seq(from=t0,to=tf,by=dt);
#
# a_jk coefficients
  A=matrix(0,nrow=nx-2,ncol=nx-1);
  for(j in 1:(nx-2)){
    for(k in 0:j){
    if (k==0){
      A[j,k+1]=(j-1)^(3-alpha)-j^(2-alpha)*(j-3+alpha);
```

```
    } else if (1 <= k && k<=j-1){
      A[j,k+1]=(j-k+1)^(3-alpha)-2*(j-k)^(3-alpha)+(j-k-1)^(3-
        alpha);
    } else
      A[j,k+1]=1;
    }
  }
#
# Initial condition
  u0=rep(0,nx-2);
  for(j in 1:(nx-2)){
    u0[j]=f(xj[j+1]);}
  ncall=0;
#
# ODE integration
  out=lsode(y=u0,times=tout,func=pde1a,
      rtol=1e-6,atol=1e-6,maxord=5);
  nrow(out)
  ncol(out)
#
# Allocate array for u(x,t)
  u=matrix(0,nt,nx);
#
# u(x,t), x ne xl,xu
  for(i in 1:nt){
    for(j in 2:(nx-1)){
      u[i,j]=out[i,j];
    }
  }
#
# Reset boundary values
  for(i in 1:nt){
   u[i,1]=ua(xl,tout[i]);
  u[i,nx]=ua(xu,tout[i]);
  }
#
# Numerical, analytical solutions, max difference
  uap=matrix(0,nt,nx);
  for(i in 1:nt){
```

```
      for(j in 1:nx){
        uap[i,j]=ua((j-1)*dx,(i-1)*dt);
      }
    max_err=max(abs(u-err));
    }
#
# Tabular numerical, analytical solutions,
# difference
    cat(sprintf("\n        t      x      u(x,t)    ua(x,t)          diff"))
      ;
    for(i in 1:nt){
    iv=seq(from=1,to=nx,by=4);
    for(j in iv){
      cat(sprintf("\n %6.2f%6.2f%10.5f%10.5f%12.3e",
        tout[i],xj[j],u[i,j],uap[i,j],u[i,j]-uap[i,j]));
    }
    cat(sprintf("\n"));
    }
#
# Plot numerical, analytical solutions
    matplot(xj,t(u),type="l",lwd=2,col="black",lty=1,
      xlab="x",ylab="u(x,t)",main="");
    matpoints(xj,t(uap),pch="o",col="black");
#
# Display maximum error
    cat(sprintf("  maximum error = %6.2e \n",max_err));
#
# Plot error at t = tf
    err_1=abs(u[nt,]-ua(xj[1:nx],tf));
    plot(xj,err_1,type="l",xlab="x",
        ylab="Max Error at t = tf",
        main="",col="black")
#
# Calls to ODE routine
    cat(sprintf("\n\n  ncall = %3d\n",ncall));
```

We can note the following details about Listing 3.1.

• Brief comments defining the test problem are followed by the deletion of previous files.

```
#
# SFPDE
#
#   ut=d(x)*(d^alpha u/dx^alpha)+p(x,t)
#
#   xl < x < xu, 0 < t < tf, xl=0, xu=1
#
#   u(x,t=0)=x^(4+alpha)
#
#   u(x=xl,t)=0; u(x=xu,t)=exp(-t)
#
#   d(x)=24/gamma(5+alpha)*x^alpha
#
#   p(x,t)=-2*exp(-t)*x^(4+alpha)
#
#   ua(x,t)=exp(-t)*x^(4+alpha)
#
# Delete previous workspaces
  rm(list=ls(all=TRUE))
```

- The ODE integrator library deSolve is accessed. Note that the setwd (set working directory) uses / rather than the usual \.

```
#
# Access functions for numerical solution
  library("deSolve");
  setwd("f:/fractional/sfpde/chap3/dirichlet");
  source("pde1a.R");
```

- The order of the fractional derivative in eq. (3.1a) is defined.

```
#
# Parameters
  alpha=1;
```

- Functions for $d(x)$ (eq. (3.1b)) and $p(x, t)$ (eq. (3.1c)) in eq. (3.1a) and the analytical solution of eq. (3.2) are defined.

```
#
# d(x)
  d=function(x) 24/gamma(5+alpha)*x^alpha;
```

```
#
# p(x,t)
  p=function(x,t) -2*exp(-t)*x^(4+alpha);
#
# Analytical solution
  ua=function(x,t) exp(-t)*x^(4+alpha);
```

- A function for IC (3.3a) is defined as the analytical solution at $t = 0$.

```
#
# Initial condition function (IC)
  f=function(x) ua(x,0);
```

- The BCs of eqs. (3.4a), (3.4b) are defined with specific values of $c_1(t), c_2(t), c_3(t), c_4(t)$ for BCs (3.3b,c).

```
#
# Boundary condition functions (BCs)
  g_0=function(t) ua(xl,t);
  g_L=function(t) ua(xu,t);
#
# Boundary condition coefficients
  c_1=function(t) 1;
  c_2=function(t) 0;
  c_3=function(t) 1;
  c_4=function(t) 0;
```

- A spatial grid of 41 points is defined for $x_l = 0 \le x \le x_u = 1$, so that xj=0,1/40,...,1.

```
#
# Spatial grid
  xl=0;xu=1;nx=41;dx=(xu-xl)/(nx-1);
  xj=seq(from=xl,to=xu,by=dx);
  cd=dx^(-alpha)/gamma(4-alpha);
```

cd is a coefficient in eq. (3.1a) that is used in the ODE routine pde1a discussed subsequently. cd is passed to pde1a without any particular specification, a feature of R.

```
#
# Independent variable for ODE integration
  t0=0;tf=1;nt=6;dt=(tf-t0)/(nt-1);
  tout=seq(from=t0,to=tf,by=dt);
```

- An interval in t of 6 points is defined for $0 \leq t \leq 1$ so that tout=0,0.2,...,1.

```
#
# Independent variable for ODE integration
  t0=0;tf=1;nt=6;dt=(tf-t0)/(nt-1);
  tout=seq(from=t0,to=tf,by=dt);
```

- The coefficients $a_{j,k}$ of eq. (1.2g) are defined with a series of ifs.

```
#
# a_jk coefficients
  A=matrix(0,nrow=nx-2,ncol=nx-1);
  for(j in 1:(nx-2)){
    for(k in 0:j){
    if (k==0){
      A[j,k+1]=(j-1)^(3-alpha)-j^(2-alpha)*(j-3+alpha);
    } else if (1 <= k && k<=j-1){
      A[j,k+1]=(j-k+1)^(3-alpha)-2*(j-k)^(3-alpha)+(j-k-1)^(3-alpha);
    } else
      A[j,k+1]=1;
    }
  }
```

- IC (3.3a) is defined.

```
#
# Initial condition
  u0=rep(0,nx-2);
  for(j in 1:(nx-2)){
    u0[j]=f(xj[j+1]);}
  ncall=0;
```

The counter for the calls to the ODE/MOL routine pde1a is also initialized.

With nx-2=41-2=39, the ICs are specified for the 39 ODEs at the interior points in x, and do not include the boundary points $x = 0, 1$ since for the latter, the dependent variables $u(x = 0, t), u(x = 1, t)$ are defined by Dirichlet BCs (3.3b,c) and not by ODEs.

The coding pertaining to nx was selected since nx is also used in the ODE/MOL routine pde1a discussed subsequently to size vectors and control fors. As explained in the subsequent discussion, the value of nx passed to pde1a is the number of ODEs, 39 (the number of ICs) plus 2 (i.e., 41).

- The system of 39 MOL/ODEs is integrated by the library integrator (available in deSolve). As expected, the inputs to lsode are the ODE function, pde1a, the IC vector u0, and the vector of output values of t, tout. The length of u0 (e.g., 39) informs lsode how many ODEs are to be integrated. func,y,times are reserved names.

```
#
# ODE integration
  out=lsode(y=u0,times=tout,func=pde1a,
      rtol=1e-6,atol=1e-6,maxord=5);
  nrow(out)
  ncol(out)
```

The numerical solution to the ODEs is returned in matrix out. In this case, out has the dimensions $nout \times (nx + 1) = 6 \times 39 + 1 = 40$, which are confirmed by the output from nrow(out),ncol(out) (included in the numerical output considered subsequently).

The offset $nx + 1$ is required since the first element of each column has the output t (also in tout), and the $2, ..., nx + 1 = 2, ..., 40$ column elements have the 39 ODE solutions.

- The solutions of the 39 ODEs returned in out by lsode are placed in u.

```
#
# Allocate array for u(x,t)
  u=matrix(0,nt,nx);
#
# u(x,t), x ne xl,xu
  for(i in 1:nt){
    for(j in 2:(nx-1)){
      u[i,j]=out[i,j];
    }
  }
```

Note that j=2,3,...,40 corresponding to the solution of the 39 ODEs.

- BCs (3.3b,c) are used to set $u(x = 0, t), u(x = 1, t)$.

```
#
# Reset boundary values
  for(i in 1:nt){
   u[i,1]=ua(xl,tout[i]);
   u[i,nx]=ua(xu,tout[i]);
   }
```

- The analytical $u(x,t)$ of eq. (3.2) is used to compute the error in the numerical solution.

```
#
# Numerical, analytical solutions, max difference
  uap=matrix(0,nt,nx);
  for(i in 1:nt){
    for(j in 1:nx){
      uap[i,j]=ua((j-1)*dx,(i-1)*dt);
    }
  max_err=max(abs(u-err));
  }
```

The absolute maximum error is determined by the `abs` and `max` utilities.

- The numerical and analytical solutions, and the error are displayed.

```
#
# Tabular numerical, analytical solutions,
# difference
  cat(sprintf("\n      t     x     u(x,t)    ua(x,t)        diff"));
  for(i in 1:nt){
  iv=seq(from=1,to=nx,by=4);
  for(j in iv){
    cat(sprintf("\n %6.2f%6.2f%10.5f%10.5f%12.3e",
      tout[i],xj[j],u[i,j],uap[i,j],u[i,j]-uap[i,j]));
  }
  cat(sprintf("\n"));
  }
```

`iv=seq(from=1,to=nx,by=4)` is used to vary the number of output lines in x, in this case, every fourth line with `by=4`.

- The numerical solution is plotted with lines (`matplot`) and the analytical solution is superimposed as points (`matpoints`).

```
#
# Plot numerical, analytical solutions
  matplot(xj,t(u),type="l",lwd=2,col="black",lty=1,
    xlab="x",ylab="u(x,t)",main="");
  matpoints(xj,t(uap),pch="o",col="black");
```

- The maximum absolute error along the solution is displayed.

```
#
# Display maximum error
  cat(sprintf("  maximum error = %6.2e \n",max_err));
```

- The absolute error at $t = t_f$ is plotted as a function of x.

```
#
# Plot error at t = tf
  err_1=abs(u[nt,]-ua(xj[1:nx],tf));
  plot(xj,err_1,type="l",xlab="x",
       ylab="Max Error at t = tf",
       main="",col="black")
```

- The numbers of calls to pde1a is displayed as a measure of the computational effort required to compute the solution.

```
#
# Calls to ODE routine
  cat(sprintf("\n\n  ncall = %3d\n",ncall));
```

The ODE/MOL routine pde1a called by lsode is discussed next.

3.2.2   SUBORDINATE ODE/MOL ROUTINE

Listing 3.2: ODE/MOL routine pde1a for eqs. (3.1) to (3.6)

```
  pde1a=function(t,u,parms){
#
# Function pde1a computes the derivative
# vector of the ODEs approximating the
# PDE
#
# Allocate the vector of the ODE
# derivatives
  nx=nx-2;
  ut=rep(0,nx);
#
# Boundary approximations of uxx
  uxx=NULL;
#
# x=0
```

```
  u0=(2*dx*g_0(t)-c_2(t)*(4*u[1]-u[2]))/
     (2*dx*c_1(t)-3*c_2(t));
  uxx_0=2*u0-5*u[1]+4*u[2]-u[3];
  uxx[1]=u[2]-2*u[1]+u0;
#
# x=1
  un=(2*dx*g_L(t)+c_4(t)*(4*u[nx]-u[nx-1]))/
     (2*dx*c_3(t)+3*c_4(t));
  uxx[nx]=un-2*u[nx]+u[nx-1];
#
# Interior approximation of uxx
  for(k in 2:(nx-1)){
    uxx[k]=u[k+1]-2*u[k]+u[k-1];
  }
#
# PDE
#
# Step through ODEs
  for(j in 1:nx){
#
#   First term in series approximation of
#   fractional derivative
    ut[j]=A[j,1]*uxx_0;
#
#   Subsequent terms in series approximation
#   of fractional derivative
    for(k in 1:j){
      ut[j]=ut[j]+A[j,k+1]*uxx[k];
#
#   Next k (next term in series)
    }
    ut[j]=cd*d(xj[j+1])*ut[j]+p(xj[j+1],t);
#
# Next j (next ODE)
  }
#
# Increment calls to pde1a
  ncall <<- ncall+1;
#
```

```
# Return derivative vector of ODEs
  return(list(c(ut)));
  }
```

We can note the following details about pde1a
  • The function is defined.

```
    pde1a=function(t,u,parms){
  #
  # Function pde1a computes the derivative
  # vector of the ODEs approximating the
  # PDE
```

t is the current value of t in eq. (3.1a). u is the 39-vector of ODE/MOL dependent variables. parm is an argument to pass parameters to pde1a (unused, but required in the argument list). The arguments must be listed in the order stated to properly interface with lsode called in the main program of Listing 3.1. The derivative vector of the LHS of eq. (3.1a) is calculated next and returned to lsode.

  • An array, ut, is allocated for the derivative $\dfrac{\partial u}{\partial t}$ of eq. (3.1a).

```
  #
  # Allocate the vector of the ODE
  # derivatives
    nx=nx-2;
    ut=rep(0,nx);
```

The value of nx is first reduced to the number of ODEs, that is, 41 to 39.

  • The integer derivative $\dfrac{\partial^2 u}{\partial x^2}$ = uxx for use in eq. (1.1) ($n = 2$) is computed.

```
  #
  # Boundary approximations of uxx
    uxx=NULL;
  #
  # x=0
    u0=(2*dx*g_0(t)-c_2(t)*(4*u[1]-u[2]))/
        (2*dx*c_1(t)-3*c_2(t));
    uxx_0=2*u0-5*u[1]+4*u[2]-u[3];
    uxx[1]=u[2]-2*u[1]+u0;
  #
```

```
# x=1
un=(2*dx*g_L(t)+c_4(t)*(4*u[nx]-u[nx-1]))/
   (2*dx*c_3(t)+3*c_4(t));
uxx[nx]=un-2*u[nx]+u[nx-1];
#
# Interior approximation of uxx
  for(k in 2:(nx-1)){
  uxx[k]=u[k+1]-2*u[k]+u[k-1];
  }
```

This code requires some additional explanation.

- uxx is nulled and then elements of this vector are added.

```
#
# Boundary approximations of uxx
  uxx=NULL;
```

- $\dfrac{\partial u(x = 0, t)}{\partial x}$ is computed with a three point FD for BC eq. (3.4a).

$$\frac{\partial u(x = 0, t)}{\partial x} \approx \frac{-3u_0 + 4u_1 - u_2}{2\Delta x} \tag{3.5a}$$

The FD approximation of BC (3.4a) is therefore

$$c_2(t)\frac{-3u_0 + 4u_1 - u_2}{2\Delta x} + c_1(t)u_0 = g_0(t) \tag{3.5b}$$

Equation (3.5b) can be solved for the boundary value $u(x = 0, t) = u_0$

$$u_0 = \frac{2\Delta x g_0(t) - c_2(t)(4u_1 - u_2)}{2\Delta x c_1(t) - 3c_2(t)} \tag{3.5c}$$

which is programmed as

```
u0=(2*dx*g_0(t)-c_2(t)*(4*u[1]-u[2]))/
   (2*dx*c_1(t)-3*c_2(t));
```

Similarly, for BC (3.4b),

$$\frac{\partial u(x = 1, t)}{\partial x} \approx \frac{3u_n - 4u_{nx} + u_{nx-1}}{2\Delta x} \tag{3.6a}$$

$$c_4(t)\frac{3u_n - 4u_{nx} + u_{nx-1}}{2\Delta x} + c_3(t)u_0 = g_L(t) \tag{3.6b}$$

$$u_n = \frac{2\Delta x g_L(t) - c_4(t)(-4u_{nx} + u_{nx-1})}{2\Delta x c_3(t) + 3c_4(t)} \tag{3.6c}$$

Equation (3.6c) is programmed as

```
un=(2*dx*g_L(t)+c_4(t)*(4*u[nx]-u[nx-1]))/
   (2*dx*c_3(t)+3*c_4(t));
```

$-\dfrac{\partial^2 u(x=0,t)}{\partial x^2}$ = uxx_0 from FD (1.2i) is programmed as

```
uxx_0=2*u0-5*u[1]+4*u[2]-u[3]
```

$-\dfrac{\partial^2 u(x=\Delta x,t)}{\partial x^2}$ = uxx[1] is programmed from FD (1.2h) as

```
uxx[1]=u[2]-2*u[1]+u0;
```

Similarly, $\dfrac{\partial^2 u(x=1-\Delta x,t)}{\partial x^2}$ = uxx[nx] is programmed from FD (1.2h) (applied to the right boundary at $x=1$) as

```
uxx[nx]=un-2*u[nx]+u[nx-1];
```

$-\dfrac{\partial^2 u(x=2\Delta x,t)}{\partial x^2}$ to $\dfrac{\partial^2 u(x=1-2\Delta x,t)}{\partial x^2}$ are programmed (from FD (1.2h)) as

```
#
# Interior approximation of uxx
  for(k in 2:(nx-1)){
    uxx[k]=u[k+1]-2*u[k]+u[k-1];
  }
```

This completes the programming of the second derivative in eq. (1.1) ($n = 2$).

- The series approximation of eqs. (1.2i), (1.2j) is computed.

```
#
# PDE
#
# Step through ODEs
  for(j in 1:nx){
#
#   First term in series approximation of
#   fractional derivative
    ut[j]=A[j,1]*uxx_0;
#
#   Subsequent terms in series approximation
```

```
#       of fractional derivative
        for(k in 1:j){
           ut[j]=ut[j]+A[j,k+1]*uxx[k];
#
#       Next k (next term in series)
        }
```

- The MOL approximation of eqs. (3.1) is programmed as

```
ut[j]=cd*d(xj[j+1])*ut[j]+p(xj[j+1],t);
```

cd is a constant defined in the main program of Listing 3.1 that includes the factor $\dfrac{1}{\Delta x^2}$ required in the preceding FD approximations of the second derivate uxx.

- After all of the MOL/ODEs are programmed (all j), the counter ncall is incremented and returned to the main program of Listing 3.1. To conclude pde1a, the MOL derivative vector ut is returned to lsode as a list for the next step along the solution in t.

```
#
# Next j (next ODE)
   }
#
# Increment calls to pde1a
   ncall <<- ncall+1;
#
# Return derivative vector of ODEs
   return(list(c(ut)));
   }
```

This completes the programming of eqs. (3.1) to (3.6). The output from the two routines of Listings 3.1, 3.2 is considered next.

3.2.3   MODEL OUTPUT

We can observe the following details about the output shown in Table 3.1.

- The solution array out is $6 \times 39 + 1 = 40$ as explained previously.

```
[1]  6
```

```
[1]  40
```

Table 3.1: Numerical solution to eqs. (3.1) to (3.6), Dirichlet BCs (*Continues.*)

[1] 6

[1] 40

t	x	u(x,t)	ua(x,t)	diff
0.00	0.00	0.00000	0.00000	0.000e+00
0.00	0.10	0.00001	0.00001	0.000e+00
0.00	0.20	0.00032	0.00032	0.000e+00
0.00	0.30	0.00243	0.00243	0.000e+00
0.00	0.40	0.01024	0.01024	0.000e+00
0.00	0.50	0.03125	0.03125	0.000e+00
0.00	0.60	0.07776	0.07776	0.000e+00
0.00	0.70	0.16807	0.16807	0.000e+00
0.00	0.80	0.32768	0.32768	0.000e+00
0.00	0.90	0.59049	0.59049	0.000e+00
0.00	1.00	1.00000	1.00000	0.000e+00
0.20	0.00	0.00000	0.00000	0.000e+00
0.20	0.10	0.00001	0.00001	1.562e-07
0.20	0.20	0.00026	0.00026	1.759e-06
0.20	0.30	0.00200	0.00199	6.250e-06
0.20	0.40	0.00840	0.00838	1.506e-05
0.20	0.50	0.02561	0.02559	2.962e-05
0.20	0.60	0.06372	0.06366	5.134e-05
0.20	0.70	0.13769	0.13760	8.159e-05
0.20	0.80	0.26840	0.26828	1.217e-04
0.20	0.90	0.48362	0.48345	1.725e-04
0.20	1.00	0.81873	0.81873	0.000e+00

```
            .                 .
            .                 .
            .                 .

     Output for t = 0.4 to 0.8 removed
            .                 .
            .                 .
            .                 .
```

Table 3.1: (*Continued.*) Numerical solution to eqs. (3.1) to (3.6), Dirichlet BCs

```
1.00   0.00    0.00000     0.00000     0.000e+00
1.00   0.10    0.00000     0.00000     8.397e-07
1.00   0.20    0.00013     0.00012     8.492e-06
1.00   0.30    0.00092     0.00089     2.975e-05
1.00   0.40    0.00384     0.00377     7.138e-05
1.00   0.50    0.01164     0.01150     1.401e-04
1.00   0.60    0.02885     0.02861     2.425e-04
1.00   0.70    0.06221     0.06183     3.850e-04
1.00   0.80    0.12109     0.12055     5.424e-04
1.00   0.90    0.21751     0.21723     2.783e-04
1.00   1.00    0.36788     0.36788     0.000e+00

maximum error = 5.48e-04

ncall = 222
```

- The numerical and analytical solutions are the same for the IC at $t = 0$ since both are defined by the analytical solution of eq. (3.2).

- The output is for $x = 0, 0.025, ..., 1$ as programmed in Listing 3.1.

- The output is for $t = 0, 0.2, ..., 1$ as programmed in Listing 3.1.

- The maximum error is 5.48e-04.

  Therefore, the error is bounded to an acceptable level with a spatial grid of 41 points.

- The computational effort is modest, ncall = 222.

- The numerical solution is the same as the solution in Table 2.1 for $\alpha = 1$ (even though the coding in Listings 2.2 and 3.2 is different for the BCs).

```
Table 2.1, alpha = 1

    1.00   0.00    0.00000     0.00000     0.000e+00
    1.00   0.10    0.00000     0.00000     8.397e-07
    1.00   0.20    0.00013     0.00012     8.492e-06
    1.00   0.30    0.00092     0.00089     2.975e-05
    1.00   0.40    0.00384     0.00377     7.138e-05
    1.00   0.50    0.01164     0.01150     1.401e-04
```

```
1.00   0.60   0.02885   0.02861   2.425e-04
1.00   0.70   0.06221   0.06183   3.850e-04
1.00   0.80   0.12109   0.12055   5.424e-04
1.00   0.90   0.21751   0.21723   2.783e-04
1.00   1.00   0.36788   0.36788   0.000e+00


Maximum error = 5.48e-04


ncall = 222


Table 3.1, alpha = 1


1.00   0.00   0.00000   0.00000   0.000e+00
1.00   0.10   0.00000   0.00000   8.397e-07
1.00   0.20   0.00013   0.00012   8.492e-06
1.00   0.30   0.00092   0.00089   2.975e-05
1.00   0.40   0.00384   0.00377   7.138e-05
1.00   0.50   0.01164   0.01150   1.401e-04
1.00   0.60   0.02885   0.02861   2.425e-04
1.00   0.70   0.06221   0.06183   3.850e-04
1.00   0.80   0.12109   0.12055   5.424e-04
1.00   0.90   0.21751   0.21723   2.783e-04
1.00   1.00   0.36788   0.36788   0.000e+00


maximum error = 5.48e-04


ncall = 222
```

Since the solution is the same as in Fig. 2.1, Chapter 2, the graphical output is not included here to conserve space.

As a second example of the use of general BCs of eqs. (3.4), eqs. (2.4) for the fractional diffusion equation are repeated here.

3.3   EXAMPLE 2, DIRICHLET BCS

$$\frac{\partial u}{\partial t} = \frac{\partial^{\alpha} u}{\partial x^{\alpha}}$$

(3.7a)

An analytical (exact) solution is not readily available for eq. (3.7a) so that only the numerical solution is presented using the same algorithms and coding validated with example 1).

The IC is

$$u(x, t = 0) = e^{-100(x-0.5)^2} \tag{3.7b}$$

a Gaussian function centered at $x = 0.5$.

Homogeneous Dirichlet BCs are selected that are consistent with the IC (3.7b).

$$u(x = 0, t) = u(x = 1, t) = 0 \tag{3.7c,d}$$

The R routines for eqs. (3.7) follow.

3.3.1 MAIN PROGRAM

Listing 3.3: Main program for eqs. (3.7)

```
#
# Fractional diffusion equation
#
#   ut=d^alpha u/dx^alpha)
#
#   xl < x < xu, 0 < t < tf, xl=0, xu=1
#
#   u(x,t=0)=e^(-c*(x-0.5)^2)
#
#   u(x=xl,t)=0; u(x=xu,t)=0
#
# Delete previous workspaces
  rm(list=ls(all=TRUE))
#
# Access functions for numerical solution
  library("deSolve");
  setwd("f:/fractional/sfpde/chap3/dirichlet");
  source("pde1b.R");
#
# Parameters
  for(ncase in 1:5){
    if(ncase==1){alpha=1;}
    if(ncase==2){alpha=1.25;}
    if(ncase==3){alpha=1.5;}
    if(ncase==4){alpha=1.75;}
    if(ncase==5){alpha=2;}
#
# Initial condition function (IC)
```

```
  f=function(x) exp(-100*(x-0.5)^2);
#
# Boundary condition functions (BCs)
  g_0=function(t) 0;
  g_L=function(t) 0;
#
# Boundary condition coefficients
  c_1=function(t) 1;
  c_2=function(t) 0;
  c_3=function(t) 1;
  c_4=function(t) 0;
#
# Spatial grid
  xl=0;xu=1;nx=51;dx=(xu-xl)/(nx-1);
  xj=seq(from=xl,to=xu,by=dx);
#
# Independent variable for ODE integration
  t0=0;tf=5;nt=6;dt=(tf-t0)/(nt-1);
  tout=seq(from=t0,to=tf,by=dt);
  ncall=0;
#
# a_jk coefficients
  A=matrix(0,nrow=nx-2,ncol=nx-1);
  for(j in 1:(nx-2)){
    for(k in 0:j){
    if (k==0){
      A[j,k+1]=(j-1)^(3-alpha)-j^(2-alpha)*(j-3+alpha);
    } else if (1 <= k && k<=j-1){
      A[j,k+1]=(j-k+1)^(3-alpha)-2*(j-k)^(3-alpha)+(j-k-1)^(3-
         alpha);
    } else
      A[j,k+1]=1;
    }
  }
#
# Initial condition
  nx=nx-2;
  u0=rep(0,nx);
  for(j in 1:nx){
```

```
      u0[j]=f(xj[j+1]);}
#
# ODE integration
  out=lsode(y=u0,times=tout,func=pde1b,
      rtol=1e-6,atol=1e-6,maxord=5);
  nrow(out)
  ncol(out)
#
# Allocate array for u(x,t)
  nx=nx+2;
  u=matrix(0,nt,nx);
#
# u(x,t), x ne xl,xu
  for(i in 1:nt){
    for(j in 2:(nx-1)){
      u[i,j]=out[i,j];
    }
  }
#
# Reset boundary values
  for(i in 1:nt){
   u[i,1]=g_0(tout[i]);
  u[i,nx]=g_L(tout[i]);
  }
#
# Tabular numerical solution
  cat(sprintf("\n\n   alpha = %4.2f\n",alpha));
  cat(sprintf("\n       t     x    u(x,t)"));
  for(i in 1:nt){
  iv=seq(from=1,to=nx,by=5);
  for(j in iv){
    cat(sprintf("\n %6.2f%6.2f%10.5f",
      tout[i],xj[j],u[i,j]));
  }
  cat(sprintf("\n"));
  }
#
# Plot numerical solution
  matplot(xj,t(u),type="l",lwd=2,col="black",lty=1,
```

```
   xlab="x",ylab="u(x,t)",main="");
#
# Calls to ODE routine
   cat(sprintf("\n   ncall = %3d\n",ncall));
#
# Next alpha (ncase)
   }
```

The main programs of Listings 3.1 and 3.3 are similar. We can note the following details.
   • Example 2 is summarized with a set of comments.

```
   #
   # Fractional diffusion equation
   #
   #   ut=d^alpha u/dx^alpha)
   #
   #   xl < x < xu, 0 < t < tf, xl=0, xu=1
   #
   #   u(x,t=0)=e^(-c*(x-0.5)^2)
   #
   #   u(x=xl,t)=0; u(x=xu,t)=0
```

   • Previous files are deleted, deSolve is accessed (for lsode), and the MOL/ODE routine is named pde1b.

```
   # Delete previous workspaces
     rm(list=ls(all=TRUE))
   #
   # Access functions for numerical solution
     library("deSolve");
     setwd("f:/fractional/sfpde/chap3/dirichlet");
     source("pde1b.R");
```

   • α is varied through the values $\alpha = 1, 1.25, 1.5, 1.75, 2$.

```
   #
   # Parameters
     for(ncase in 1:5){
        if(ncase==1){alpha=1;}
        if(ncase==2){alpha=1.25;}
        if(ncase==3){alpha=1.5;}
```

```
         if(ncase==4){alpha=1.75;}
         if(ncase==5){alpha=2;}
```

- IC (3.7b) is defined (a Gaussian function centered around $x = 0.5$.

```
#
# Initial condition function (IC)
  f=function(x) exp(-100*(x-0.5)^2);
```

- The homogeneous Dirichlet BC functions of eqs. (3.7c,d) are defined.

```
#
# Boundary condition functions (BCs)
  g_0=function(t) 0;
  g_L=function(t) 0;
```

- The coefficients in the general BCs (3.4) are defined for Dirichlet BCs.

```
#
# Boundary condition coefficients
  c_1=function(t) 1;
  c_2=function(t) 0;
  c_3=function(t) 1;
  c_4=function(t) 0;
```

- The spatial grid has 51 points for the interval $0 \le x \le 1$, and with the values $xj = 0, 0.02, ..., 1$.

```
#
# Spatial grid
  xl=0;xu=1;nx=51;dx=(xu-xl)/(nx-1);
  xj=seq(from=xl,to=xu,by=dx);
```

The number of points in x was selected to give good spatial resolution of the solutions with a sharp variation in x, e.g., for $\alpha = 1$ as explained subsquently.

- The interval in t is $0 \le t \le 5$ with 6 output points so $tout = 0, 1, ..., 5$. The number of output points in t was selected to indicate the t evolution of the solution without the non-zero part of the solution reaching the left and right boundaries, $x = 0, 1$ in violation of the homogeneous Dirichlet BCs (3.7c,d).

```
#
# Independent variable for ODE integration
  t0=0;tf=5;nt=6;dt=(tf-t0)/(nt-1);
  tout=seq(from=t0,to=tf,by=dt);
  ncall=0;
```

The integer counter for the calls to pde1b is initialzed.

- The A matrix of eq. (1.2g) is defined.

- IC (3.7b) is specified.

```
#
# Initial condition
  nx=nx-2;
  u0=rep(0,nx);
  for(j in 1:nx){
    u0[j]=f(xj[j+1]);}
```

nx is first reduced from 51 to 49, the number of MOL/ODEs at the interior grid points in x. This value (49) is then passed to the ODE/MOL routine pde1b.

- The 49 ODEs are integrated by lsode which is informed of the number of ODEs through the length of the IC vector u0.

```
#
# ODE integration
  out=lsode(y=u0,times=tout,func=pde1b,
      rtol=1e-6,atol=1e-6,maxord=5);
  nrow(out)
  ncol(out)
```

The dimensions of the solution matrix out are $6 \times 49 + 1 = 50$ as explained previously and confirmed by the output discussed subsequently. The input and output parameters for lsode are explained in the preceding discussion of Listing 3.1.

- The numerical solution is placed in array u. The nx=51 values include the boundary values at $x = 0, 1$.

```
#
# Allocate array for u(x,t)
  nx=nx+2;
```

```
  u=matrix(0,nt,nx);
#
# u(x,t), x ne xl,xu
  for(i in 1:nt){
    for(j in 2:(nx-1)){
      u[i,j]=out[i,j];
    }
  }
```

The subscripting is explained in the preceding discussion of Listing 3.1.

- The boundary values at $x = 0, 1$ are reset.

```
#
# Reset boundary values
  for(i in 1:nt){
   u[i,1]=g_0(tout[i]);
  u[i,nx]=g_L(tout[i]);
  }
```

- Abbreviated numerical values of the solution are displayed (every fifth value in x from by=5).

```
#
# Tabular numerical solution
  cat(sprintf("\n\n   alpha = %4.2f\n",alpha));
  cat(sprintf("\n       t     x     u(x,t)"));
  for(i in 1:nt){
  iv=seq(from=1,to=nx,by=5);
  for(j in iv){
    cat(sprintf("\n %6.2f%6.2f%10.5f",
      tout[i],xj[j],u[i,j]));
  }
  cat(sprintf("\n"));
  }
```

An analytical solution is not readily available to add to this output.

- The numerical solution is plotted with matplot.

```
#
```

```
# Plot numerical solution
  matplot(xj,t(u),type="l",lwd=2,col="black",lty=1,
    xlab="x",ylab="u(x,t)",main="");
```

A transpose is required, t(u), so that the number of rows of u equals the number of elements in xj.

- The calls to ode1b is displayed for the complete solution, and the next case (next α) is initiated.

```
#
# Calls to ODE routine
  cat(sprintf("\n   ncall = %3d\n",ncall));
#
# Next alpha (ncase)
  }
```

3.3.2   SUBORDINATE ODE/MOL ROUTINE

pde1b is the same as pde1a of Listing 3.2 except for

- The name, pde1b in place of pde1a.

- The programming of eq. (3.7a).

```
      for(k in 1:j){
        ut[j]=ut[j]+A[j,k+1]*uxx[k];
#
#   Next k (next term in series)
      }
#
# Next j (next ODE)
  }
```

Equation (3.7a) has only a unit coefficient multiplying the fractional derivative and no inhomogeneous term.

The output for $\alpha = 1$ is the same as from Table 2.2.

3.3.3   MODEL OUTPUT

Numerical solution to eqs. (3.7), Dirichlet BCs, $\alpha = 1$ is shown in Table 3.2.

Similarly, the solutions for $\alpha = 1.25, 1.5, 1.75, 2$ are the same as in Table 2.2.

Since the solution in Table 3.2 has already been plotted in Fig. 2.5, it is not repeated here.

The following discussion pertains to example 2 with Neumann and Robin BCs.

3.4 EXAMPLE 2, NEUMANN BCS

The changes from Dirichlet to Neumann BCs using the general BCs (3.4) are considered next.

3.4.1 MAIN PROGRAM

The main program is the same as in Listing 3.1 except the coefficients in BCs (3.4) are changed.

```
Dirichlet BCs (Listing 3.1)

#
# Boundary condition coefficients
  c_1=function(t) 1;
  c_2=function(t) 0;
  c_3=function(t) 1;
  c_4=function(t) 0;


Neumann BCs

#
# Boundary condition coefficients
  c_1=function(t) 0;
  c_2=function(t) 1;
  c_3=function(t) 0;
  c_4=function(t) 1;
```

3.4.2 SUBORDINATE ODE/MOL ROUTINE

The MOL/ODE routine is the same as in Listing 3.2 (different values of $c_1(t)$ to $c_4(t)$ are used as listed above).

Table 3.2: Numerical solution to eqs. (3.7) to Dirichlet BCs, $\alpha = 1$ (*Continues.*)

```
alpha = 1.00

    t     x     u(x,t)
 0.00  0.00   0.00000
 0.00  0.10   0.00000
 0.00  0.20   0.00012
 0.00  0.30   0.01832
 0.00  0.40   0.36788
 0.00  0.50   1.00000
 0.00  0.60   0.36788
 0.00  0.70   0.01832
 0.00  0.80   0.00012
 0.00  0.90   0.00000
 0.00  1.00   0.00000

 1.00  0.00   0.00000
 1.00  0.10   0.00001
 1.00  0.20   0.00152
 1.00  0.30   0.07996
 1.00  0.40   0.68783
 1.00  0.50   0.86165
 1.00  0.60   0.13897
 1.00  0.70   0.00246
 1.00  0.80   0.00000
 1.00  0.90  -0.00000
 1.00  1.00   0.00000

           .
           .
           .

Output for t = 2
  to 4 removed

           .
           .
           .
```

Table 3.2: (*Continued.*) Numerical solution to eqs. (3.7) to Dirichlet BCs, $\alpha = 1$

```
5.00   0.00    0.00000
5.00   0.10    0.02636
5.00   0.20    0.35216
5.00   0.30    0.98993
5.00   0.40    0.39407
5.00   0.50    0.00612
5.00   0.60   -0.00061
5.00   0.70   -0.00061
5.00   0.80   -0.00061
5.00   0.90   -0.00061
5.00   1.00    0.00000


ncall = 345
```

3.4.3   MODEL OUTPUT

The output for $\alpha = 1, t = 5$, for Neumann BCs (Table 3.3) is similar to Dirichlet BC (Table 3.2). This is also demonstrated by Fig. 3.1 which is similar to Fig. 2.5

```
Table 3.2, Dirichlet

  alpha = 1.00

  5.00   0.00    0.00000
  5.00   0.10    0.02636
  5.00   0.20    0.35216
  5.00   0.30    0.98993
  5.00   0.40    0.39407
  5.00   0.50    0.00612
  5.00   0.60   -0.00061
  5.00   0.70   -0.00061
  5.00   0.80   -0.00061
  5.00   0.90   -0.00061
  5.00   1.00    0.00000


ncall = 345
```

```
Table 3.3, Neumann

  alpha = 1.00

  5.00   0.00    0.00000
  5.00   0.10    0.02649
  5.00   0.20    0.35228
  5.00   0.30    0.99006
  5.00   0.40    0.39419
  5.00   0.50    0.00624
  5.00   0.60   -0.00049
  5.00   0.70   -0.00049
  5.00   0.80   -0.00049
  5.00   0.90   -0.00049
  5.00   1.00    0.00000


  ncall = 345
```

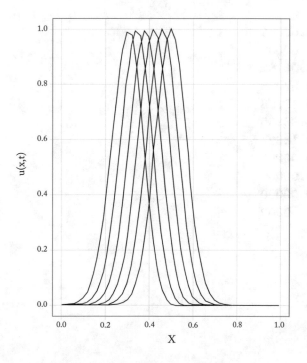

Figure 3.1: Numerical solution of eqs. (3.7), Neumann BCs, $\alpha = 1$.

Table 3.3: Numerical solution to eqs. (3.7), Neumann BCs, $\alpha = 1$ (*Continues.*)

```
alpha = 1.00

    t      x      u(x,t)
 0.00   0.00    0.00000
 0.00   0.10    0.00000
 0.00   0.20    0.00012
 0.00   0.30    0.01832
 0.00   0.40    0.36788
 0.00   0.50    1.00000
 0.00   0.60    0.36788
 0.00   0.70    0.01832
 0.00   0.80    0.00012
 0.00   0.90    0.00000
 0.00   1.00    0.00000

 1.00   0.00    0.00000
 1.00   0.10    0.00001
 1.00   0.20    0.00152
 1.00   0.30    0.07996
 1.00   0.40    0.68783
 1.00   0.50    0.86165
 1.00   0.60    0.13897
 1.00   0.70    0.00246
 1.00   0.80    0.00000
 1.00   0.90   -0.00000
 1.00   1.00    0.00000
                 .
                 .
                 .
Output for t = 2
   to 4 removed
                 .
                 .
                 .
```

Table 3.3: (*Continued.*) Numerical solution to eqs. (3.7), Neumann BCs, $\alpha = 1$ Neumann BCs, $\alpha = 1$

```
5.00  0.00    0.00000
5.00  0.10    0.02649
5.00  0.20    0.35228
5.00  0.30    0.99006
5.00  0.40    0.39419
5.00  0.50    0.00624
5.00  0.60   -0.00049
5.00  0.70   -0.00049
5.00  0.80   -0.00049
5.00  0.90   -0.00049
5.00  1.00    0.00000

ncall = 345
```

The slope of the solution near the boundaries at $x = 0, 1$ is close to zero indicating homogeneous Neumann BCs are closely approximated.

Also, for $\alpha = 1$, eqs. (3.7) correspond to the linear advection equation, eqs. (2.5), so that the solution is a translation of the IC right to left in t. This is clear from Figs. 2.5 and 3.1.

To conclude, eqs. (3.4) are considered for Robin BCs, with BCs (3.4), $g_0(t) = g_L(t) = 0$ in eqs. (3.3b,c).

3.5   EXAMPLE 2, ROBIN BCS

The changes from Dirichlet to Robin BCs using the general BCs (3.4), are straightforward.

3.5.1   MAIN PROGRAM

The main program is the same as in Listing 3.1 except the coefficients in BCs (3.4), are changed.

```
Dirichlet BCs (Listing 3.1)

#
# Boundary condition coefficients
  c_1=function(t) 1;
  c_2=function(t) 0;
  c_3=function(t) 1;
  c_4=function(t) 0;
```

Robin BCs

```
#
# Boundary condition coefficients
  c_1=function(t) -1;
  c_2=function(t)  1;
  c_3=function(t)  1;
  c_4=function(t)  1;
```

These coefficients reflect Robin BCs that are frequently used in applications. For example,

$$D\frac{\partial u(x = x_l, t)}{\partial x} + k(u_a(t) - u(x = x_l, t)) = 0$$

$$D\frac{\partial u(x = x_u, t)}{\partial x} + k(u(x = x_u, t) - u_a(t)) = 0$$

where $D > 0$ is a diffusivity, $k > 0$ is a transfer coefficient and $u_a(t) > 0$ is an ambient condition.

3.5.2   SUBORDINATE ODE/MOL ROUTINE

The MOL/ODE routine is the same as in Listing 3.2 (different values of $c_1(t)$ to $c_4(t)$ are used as listed above).

3.5.3   MODEL OUTPUT

Abbreviated numerical output and the graphical output for $\alpha = 1$ are shown in Table 3.4 and Fig. 3.2, respectively.

The output for $\alpha = 1, t = 5$, Dirichlet (Table 3.2), Neumann (Table 3.3) and Robin (Table 3.4) is similar. This is also demonstrated by Fig. 3.2 which is similar to Figs. 2.5 and 3.1

```
Table 3.2, Dirichlet

   alpha = 1.00

   5.00   0.00    0.00000
   5.00   0.10    0.02636
   5.00   0.20    0.35216
   5.00   0.30    0.98993
   5.00   0.40    0.39407
   5.00   0.50    0.00612
   5.00   0.60   -0.00061
```

```
5.00   0.70   -0.00061
5.00   0.80   -0.00061
5.00   0.90   -0.00061
5.00   1.00    0.00000

ncall = 345
```

Table 3.3, Neumann

```
alpha = 1.00

5.00   0.00    0.00000
5.00   0.10    0.02649
5.00   0.20    0.35228
5.00   0.30    0.99006
5.00   0.40    0.39419
5.00   0.50    0.00624
5.00   0.60   -0.00049
5.00   0.70   -0.00049
5.00   0.80   -0.00049
5.00   0.90   -0.00049
5.00   1.00    0.00000

ncall = 345
```

Table 3.4, Robin

```
alpha = 1.00

5.00   0.00    0.00000
5.00   0.10    0.02649
5.00   0.20    0.35228
5.00   0.30    0.99006
5.00   0.40    0.39419
5.00   0.50    0.00624
5.00   0.60   -0.00049
5.00   0.70   -0.00049
```

```
5.00   0.80   -0.00049
5.00   0.90   -0.00049
5.00   1.00    0.00000

ncall = 345
```

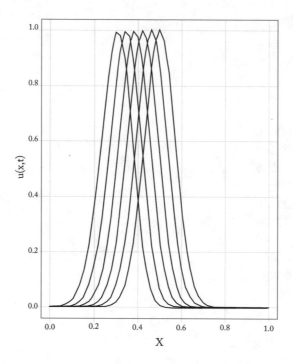

Figure 3.2: Numerical solution of eqs. (3.7), Robin BCs, $\alpha = 1$.

The fractional diffusion equation eq. (3.7a) with $\alpha = 1$ is a stringent test of the numerical procedures for SFPDEs as reflected in the traveling wave solution for a narrow Gaussian function, IC (3.7b). That is, the solution retains it form with increasing t.

Table 3.4: Numerical solution to eqs. (3.7), Robin BCs, $\alpha = 1$ (*Continues.*)

```
alpha = 1.00

    t     x     u(x,t)
  0.00  0.00   0.00000
  0.00  0.10   0.00000
  0.00  0.20   0.00012
  0.00  0.30   0.01832
  0.00  0.40   0.36788
  0.00  0.50   1.00000
  0.00  0.60   0.36788
  0.00  0.70   0.01832
  0.00  0.80   0.00012
  0.00  0.90   0.00000
  0.00  1.00   0.00000

  1.00  0.00   0.00000
  1.00  0.10   0.00001
  1.00  0.20   0.00152
  1.00  0.30   0.07996
  1.00  0.40   0.68783
  1.00  0.50   0.86165
  1.00  0.60   0.13897
  1.00  0.70   0.00246
  1.00  0.80   0.00000
  1.00  0.90  -0.00000
  1.00  1.00   0.00000
              .
              .
              .
Output for t = 2
  to 4 removed
              .
              .
              .
```

Table 3.4: (*Continued.*) Numerical solution to eqs. (3.7), Robin BCs, $\alpha = 1$

```
5.00   0.00    0.00000
5.00   0.10    0.02649
5.00   0.20    0.35228
5.00   0.30    0.99006
5.00   0.40    0.39419
5.00   0.50    0.00624
5.00   0.60   -0.00049
5.00   0.70   -0.00049
5.00   0.80   -0.00049
5.00   0.90   -0.00049
5.00   1.00    0.00000

ncall = 345
```

3.6   SUMMARY AND CONCLUSIONS

In this chapter, application of the MOL approach to SFPDEs with Dirichlet, Neumann and Robin BCs is demonstrated with the Caputo fractional derivative used in all cases. The discussion is in terms of two examples, with and without an exact solution. For the latter, variation in the number of spatial points (not explicitly reported in the discussion) and a comparison with the limiting case of integer PDEs ($\alpha = 1, 2$) infers valid numerical solutions.

Consideration is next given to additional types of SFPDE systems through the use of the procedures discussed in this chapter.

REFERENCES

[1] Sousa, E. (2011), Numerical approximations for fractional diffusion equations via splines, *Computers and Mathematics with Applications*, 62, 938–944. 65

CHAPTER 4

Convection SFPDEs

4.1 INTRODUCTION

In this chapter, the advection SFPDEs of eqs. (2.4) and (2.5) are considered again through the addition of an integer convection term to eq. (2.4a),

$$\frac{\partial u}{\partial t} = -v\frac{\partial u}{\partial x} + \frac{\partial^\alpha u}{\partial x^\alpha} \qquad (4.1a)$$

$$u(x, t = 0) = e^{-100(x-0.5)^2} \qquad (4.1b)$$

$$\frac{\partial u(x = 0, t)}{\partial x} = \frac{\partial u(x = 1, t)}{\partial x} = 0 \qquad (4.1c,d)$$

The first RHS term of eq. (4.1a), RHS1 $= -v\dfrac{\partial u}{\partial x}$, is considered an integer convection term approximated with finite differences (FDs). The second RHS term, RHS2 $= \dfrac{\partial^\alpha u}{\partial x^\alpha}$, is a fractional derivative which for $\alpha = 1$ corresponds to convection, and is approximated by the algorithm of eqs. (1.2). In particular, various combinations of the two RHS terms provide a comparison of integer and fractional convection.

With RHS2 = 0, eq. (4.1a) is the linear advection equation, eq. (2.5a), with the traveling wave solution of eqs. (2.5b), (2.5c). Eqs. (2.5) constitute a stringent test problem that provides some interesting comparisons of the contribuions of RHS1, RHS2.

An analytical solution is not readily available, but the numerical methods in Chapter 3 that were verified with analytical solution (3.2) are used here.

4.2 INTEGER/FRACTIONAL CONVECTION MODEL

The programming of eqs. (4.1) is considered next, starting with a main program.

4.2.1 MAIN PROGRAM

Listing 4.1: Main program for eqs. (4.1)

```
#
# Convection SFPDE
```

```
#
#    ut=-v*ux + (d^alpha u/dx^alpha)
#
#    xl < x < xu, 0 < t < tf, xl=0, xu=1
#
#    u(x,t=0)=e^(-100*(x-0.5)^2)
#
#    c2(t)*ux(x=xl,t)+c1(t)*u(x=xl,t)=0
#
#    c4(t)*ux(x=xu,t)+c3(t)*u(x=xu,t)=0
#
# Delete previous workspaces
  rm(list=ls(all=TRUE))
#
# Access functions for numerical solution
  library("deSolve");
  setwd("f:/fractional/sfpde/chap4");
  source("pde1a.R");
#
# Parameters
  alpha=1;
  v=0;
#
# Initial condition function (IC)
  f=function(x) exp(-100*(x-0.5)^2);
#
# Boundary condition functions
  g_0=function(t) 0;
  g_L=function(t) 0;
#
# Boundary condition coefficients
  c_1=function(t) 0;
  c_2=function(t) 1;
  c_3=function(t) 0;
  c_4=function(t) 1;
#
# Spatial grid
  xl=0;xu=1;nx=51;dx=(xu-xl)/(nx-1);
  xj=seq(from=xl,to=xu,by=dx);
```

```
  cd=dx^(-alpha)/gamma(4-alpha);
#
# Independent variable for ODE integration
  t0=0;tf=0.2;nt=5;dt=(tf-t0)/(nt-1);
  tout=seq(from=t0,to=tf,by=dt);
#
# a_jk coefficients
  A=matrix(0,nrow=nx-2,ncol=nx-1);
  for(j in 1:(nx-2)){
    for(k in 0:j){
    if (k==0){
      A[j,k+1]=(j-1)^(3-alpha)-j^(2-alpha)*(j-3+alpha);
    } else if (1 <= k && k<=j-1){
      A[j,k+1]=(j-k+1)^(3-alpha)-2*(j-k)^(3-alpha)+(j-k-1)^(3-
          alpha);
    } else
      A[j,k+1]=1;
    }
  }
#
# Initial condition
  u0=rep(0,nx-2);
  for(j in 1:(nx-2)){
    u0[j]=f(xj[j+1]);}
  ncall=0;
#
# ODE integration
  out=lsode(y=u0,times=tout,func=pde1a,
      rtol=1e-6,atol=1e-6,maxord=5);
  nrow(out)
  ncol(out)
#
# Allocate array for u(x,t)
  u=matrix(0,nt,nx);
#
# u(x,t), x ne xl,xu
  for(i in 1:nt){
    for(j in 2:(nx-1)){
      u[i,j]=out[i,j];
```

```
      }
    }
#
# Reset boundary values
  for(i in 1:nt){
   u[i,1]=g_0(tout[i]);
   u[i,nx]=g_L(tout[i]);
   }
#
# Display parameters
  cat(sprintf("\n alpha = %5.2f  v = %5.2f",alpha,v));
#
# Tabular numerical solution
  cat(sprintf("\n       t      x     u(x,t)"));
  for(i in 1:nt){
  iv=seq(from=1,to=nx,by=5);
  for(j in iv){
    cat(sprintf("\n %6.2f%6.2f%10.5f",
      tout[i],xj[j],u[i,j]));
  }
  cat(sprintf("\n"));
  }
#
# Plot numerical solution
  matplot(xj,t(u),type="l",lwd=2,col="black",lty=1,
    xlab="x",ylab="u(x,t)",main="");
#
# Calls to ODE routine
  cat(sprintf("\n\n  ncall = %3d\n",ncall));
```

We can note the following details of Listing 4.1.

• Brief comments defining the test problem are followed by the deletion of previous files.

```
    #
    # Convection SFPDE
    #
    #   ut=-v*ux + (d^alpha u/dx^alpha)
    #
    #   xl < x < xu, 0 < t < tf, xl=0, xu=1
    #
```

```
#    u(x,t=0)=e^(-100*(x-0.5)^2)
#
#    c2(t)*ux(x=xl,t)+c1(t)*u(x=xl,t)=0
#
#    c4(t)*ux(x=xu,t)+c3(t)*u(x=xu,t)=0
#
# Delete previous workspaces
  rm(list=ls(all=TRUE))
```

- The ODE integrator library deSolve is accessed. pde1a is the ODE/MOL routine (discussed subsequently).

```
#
# Access functions for numerical solution
  library("deSolve");
  setwd("f:/fractional/sfpde/chap4");
  source("pde1a.R");
```

- The parameters v, α in eq. (4.1a) are set numerically. Of particular interest in the following analysis is the relative effect of the linear convection (via v) and the apparent fractional convection (via α) from eq. (4.1a).

```
#
# Parameters
  alpha=1;
  v=0;
```

- The Gaussian IC function of eq. (4.1b) is defined.

```
#
# Initial condition function (IC)
  f=function(x) exp(-100*(x-0.5)^2);
```

- The RHS functions and the coefficients of eqs. (3.4) are defined for Neumann BCs.

```
#
# Boundary condition functions
  g_0=function(t) 0;
  g_L=function(t) 0;
#
# Boundary condition coefficients
```

```
c_1=function(t) 0;
c_2=function(t) 1;
c_3=function(t) 0;
c_4=function(t) 1;
```

- A spatial grid of 51 points is defined for $x_l = 0 \leq x \leq x_u = 1$, so that xj=0,1/50,...,1.

```
#
# Spatial grid
  xl=0;xu=1;nx=51;dx=(xu-xl)/(nx-1);
  xj=seq(from=xl,to=xu,by=dx);
  cd=dx^(-alpha)/gamma(4-alpha);
```

cd is a constant used in the ODE/MOL routine pde1a.

- An interval in t of 5 points is defined for $0 \leq t \leq 0.2$ so that tout=0,0.05,...,0.2.

```
#
# Independent variable for ODE integration
  t0=0;tf=0.2;nt=5;dt=(tf-t0)/(nt-1);
  tout=seq(from=t0,to=tf,by=dt);
```

- The A matrix of eq. (1.2g) is defined.

- The IC (4.1b) is set numerically.

```
#
# Initial condition
  u0=rep(0,nx-2);
  for(j in 1:(nx-2)){
    u0[j]=f(xj[j+1]);}
  ncall=0;
```

The IC is for the 49 interior ODEs with the derivatives in t computed in pde1a.

- The 49 ODEs are integrated by lsode which is informed of the number of ODEs through the length of the IC vector u0.

```
#
# ODE integration
  out=lsode(y=u0,times=tout,func=pde1a,
      rtol=1e-6,atol=1e-6,maxord=5);
  nrow(out)
  ncol(out)
```

The dimensions of the solution matrix out are $6 \times 49 + 1 = 50$ (the offset +1 includes the value of t in each solution vector of 49 elements.

The input parameters for lsode are the IC vector u0, the vector of output values of t, tout, the ODE/PDE routine pde1a, the error tolerances rtol,atol (the default values are also 1e-06), and the maximum order of the ODE integration algorithm in lsode. y,times,func,rtol,atol,maxord are reserved names and therefore the arguments can be listed in any order.

The numerical solution of the 49 ODEs is in out. nrow,ncol give the dimensions of out, 6×50, as confirmed in the output discussed subsequently.

- The solution matrix out is placed in matrix u with dimensions 6×51 (so that it includes the boundary values of the solution at $x = x_l = 0, x = x_u = 1$).

```
#
# Allocate array for u(x,t)
  u=matrix(0,nt,nx);
#
# u(x,t), x ne xl,xu
  for(i in 1:nt){
    for(j in 2:(nx-1)){
      u[i,j]=out[i,j];
    }
  }
```

- The boundary values are reset according to eqs. (4.1c,d) (these values are not returned by lsode since they do not result from the integration of ODEs).

```
#
# Reset boundary values
  for(i in 1:nt){
   u[i,1]=g_0(tout[i]);
   u[i,nx]=g_L(tout[i]);
   }
```

- v, α are displayed, followed by the numerical solution as a function of x and t (for every fifth value of x from by=5).

```
#
# Display parameters
  cat(sprintf("\n alpha = %5.2f   v = %5.2f",alpha,v));
```

```
#
# Tabular numerical solution
  cat(sprintf("\n        t      x     u(x,t)"));
  for(i in 1:nt){
  iv=seq(from=1,to=nx,by=5);
  for(j in iv){
    cat(sprintf("\n %6.2f%6.2f%10.5f",
      tout[i],xj[j],u[i,j]));
  }
  cat(sprintf("\n"));
  }
```

- $u(x,t)$ is plotted parametically against t. The transpose t(u) is used so that the dimensions of u and xj agree (i.e., 51).

```
#
# Plot numerical solution
  matplot(xj,t(u),type="l",lwd=2,col="black",lty=1,
    xlab="x",ylab="u(x,t)",main="");
```

- The number of calls to pde1a is displayed at the end of the solution.

```
#
# Calls to ODE routine
  cat(sprintf("\n\n  ncall = %3d\n",ncall));
```

The ODE/MOL routine is next.

4.2.2   SUBORDINATE ODE/MOL ROUTINE

pde1a called by lsode in Listing 4.1 follows.

Listing 4.2: ODE/MOL routine for eqs. (4.1)

```
  pde1a=function(t,u,parms){
#
# Function pde1a computes the derivative
# vector of the ODEs approximating the
# PDE
#
# Allocate the vector of the ODE
# derivatives
```

```
  nx=nx-2;
  ut=rep(0,nx);
#
# Array for ux
  ux=rep(0,nx);
  for(j in 1:nx){
    if(j==1){ux[1]=(u[1]-g_0(t))/dx};
    if(j>1) {ux[j]=(u[j]-u[j-1])/dx};
  }
#
# Boundary approximations of uxx
  uxx=NULL;
#
# x=0
  u0=(2*dx*g_0(t)-c_2(t)*(4*u[1]-u[2]))/
     (2*dx*c_1(t)-3*c_2(t));
  uxx_0=2*u0-5*u[1]+4*u[2]-u[3];
  uxx[1]=u[2]-2*u[1]+u0;
#
# x=1
  un=(2*dx*g_L(t)+c_4(t)*(4*u[nx]-u[nx-1]))/
     (2*dx*c_3(t)+3*c_4(t));
  uxx[nx]=un-2*u[nx]+u[nx-1];
#
# Interior approximation of uxx
  for(k in 2:(nx-1)){
    uxx[k]=u[k+1]-2*u[k]+u[k-1];
  }
#
# PDE
#
# Step through ODEs
  for(j in 1:nx){
#
#   First term in series approximation of
#   fractional derivative
    ut[j]=A[j,1]*uxx_0;
#
#   Subsequent terms in series approximation
```

```
#    of fractional derivative
     for(k in 1:j){
       ut[j]=ut[j]+A[j,k+1]*uxx[k];
#
#    Next k (next term in series)
     }
     ut[j]=-v*ux[j]+cd*ut[j];
#
# Next j (next ODE)
   }
#
# Increment calls to pde1a
  ncall <<- ncall+1;
#
# Return derivative vector of ODEs
  return(list(c(ut)));
  }
```

We can note the following details about Listing 4.1.

- The function is defined. t is the current value of t. u is the current solution vector of eq. (4.1a) for the 49 interior points in x. parm for passing parameters to pde1a is unusued, but must be included as the third parameter.

```
  pde1a=function(t,u,parms){
#
# Function pde1a computes the derivative
# vector of the ODEs approximating the
# PDE
```

- Array ut is declared (allocated) for the $51 - 2 = 49$ ODE derivatives in t.

```
#
# Allocate the vector of the ODE
# derivatives
  nx=nx-2;
  ut=rep(0,nx);
```

- $\dfrac{\partial u(x,t)}{\partial x}$ in eq. (4.1a) is computed by first order finite differences (FDs).

```
#
```

```
# Array for ux
  ux=rep(0,nx);
  for(j in 1:nx){
    if(j==1){ux[1]=(u[1]-g_0(t))/dx};
    if(j>1) {ux[j]=(u[j]-u[j-1])/dx};
  }
```

$\dfrac{\partial^2 u(x=0,t)}{\partial x^2}$, $\dfrac{\partial^2 u(x=1,t)}{\partial x^2}$ are computed by FD (3.5c) for $x=0$ and a corresponding FD at $x=1$.

```
#
# Boundary approximations of uxx
  uxx=NULL;
#
# x=0
  u0=(2*dx*g_0(t)-c_2(t)*(4*u[1]-u[2]))/
     (2*dx*c_1(t)-3*c_2(t));
  uxx_0=2*u0-5*u[1]+4*u[2]-u[3];
  uxx[1]=u[2]-2*u[1]+u0;
#
# x=1
  un=(2*dx*g_L(t)+c_4(t)*(4*u[nx]-u[nx-1]))/
     (2*dx*c_3(t)+3*c_4(t));
  uxx[nx]=un-2*u[nx]+u[nx-1];
```

- $\dfrac{\partial^2 u(x,t)}{\partial x^2}$ is computed by FD (1.2h).

```
#
# Interior approximation of uxx
  for(k in 2:(nx-1)){
    uxx[k]=u[k+1]-2*u[k]+u[k-1];
  }
```

- The fractional derivative of eq. (4.1a) is computed with eq. (1.2j).

```
#
# PDE
#
# Step through ODEs
  for(j in 1:nx){
```

```
#
#    First term in series approximation of
#    fractional derivative
     ut[j]=A[j,1]*uxx_0;
#
#    Subsequent terms in series approximation
#    of fractional derivative
     for(k in 1:j){
        ut[j]=ut[j]+A[j,k+1]*uxx[k];
#
#    Next k (next term in series)
     }
```

- Equation (4.1a) is programmed at the interior points in x.

```
     ut[j]=-v*ux[j]+cd*ut[j];
#
# Next j (next ODE)
   }
```

- The counter for the calls to pde1a is incremented.

```
#
# Increment calls to pde1a
  ncall <<- ncall+1;
```

- The derivative vector ut for the LHS of eq. (4.1a) is returned to lsode as a list (as required by lsode).

```
#
# Return derivative vector of ODEs
  return(list(c(ut)));
  }
```

The final } concludes pde1a.

The output from the main program and ODE/MOL routines of Listings 4.1 and 4.2 is considered next.

4.2.3 SFPDE OUTPUT

Abbreviated numerical output for eqs. (4.1) is shown in Table 4.1.

We can note the following details about the output in Table 4.1.

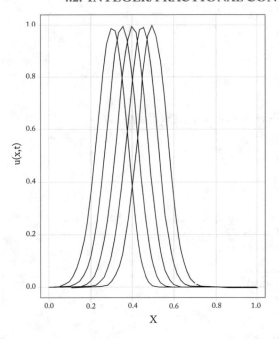

Figure 4.1: Numerical solution of eqs. (4.1), $v = 0, \alpha = 1$.

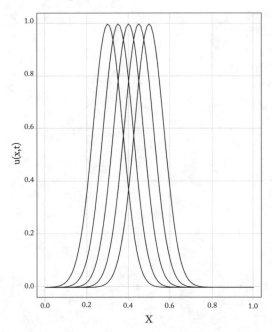

Figure 4.2: Numerical solution of eqs. (4.1), $v = 0, \alpha = 1, nx = 101$.

Table 4.1: Numerical solution to eqs. (4.1) (*Continues.*)

```
[1] 5

[1] 50

 alpha =  1.00  v =  0.00

     t      x     u(x,t)
  0.00   0.00   0.00000
  0.00   0.10   0.00000
  0.00   0.20   0.00012
  0.00   0.30   0.01832
  0.00   0.40   0.36788
  0.00   0.50   1.00000
  0.00   0.60   0.36788
  0.00   0.70   0.01832
  0.00   0.80   0.00012
  0.00   0.90   0.00000
  0.00   1.00   0.00000
                  .
                  .
                  .

 Output for t = 0.05,
    0.1,0.15 removed
                  .
                  .
                  .
```

Table 4.1: (*Continued.*) Numerical solution to eqs. (4.1)

```
0.20   0.00    0.00000
0.20   0.10    0.02649
0.20   0.20    0.35228
0.20   0.30    0.99006
0.20   0.40    0.39419
0.20   0.50    0.00624
0.20   0.60   -0.00049
0.20   0.70   -0.00049
0.20   0.80   -0.00049
0.20   0.90   -0.00049
0.20   1.00    0.00000

ncall = 345
```

- The dimensions of the output matrix out are $6 \times 49 + 1 = 50$ as discussed previously.

  ```
  [1] 6
  ```

  ```
  [1] 50
  ```

- The solution is for fractional convection only.

  ```
  alpha =  1.00  v =  0.00
  ```

- The maximum in the Gaussian IC is at $x = 0.5$.

  ```
  0.00   0.40    0.36788
  0.00   0.50    1.00000
  0.00   0.60    0.36788
  ```

- The maxiumum has moved to approximately $x = 0.3$ at $t = 0.2$ so the fractional effective velocity is negative.

  ```
  0.20   0.20    0.35228
  0.20   0.30    0.99006
  0.20   0.40    0.39419
  ```

- The computational effort is modest.

```
ncall = 345
```

The graphical output is in Fig. 4.1.

Figure 4.1 indicates the IC Gaussian pulse has moved right to left. The lack of numerical distortion (e.g., diffusion, oscillation) is particularly noteworthy.

Figure 4.1 suggests that $nx = 51$ may have introduced some spatial gridding effects (lack of smoothness). To investigate this point, the output for $nx = 101$ is shown in Table 4.2.

The maximum in the solution has moved from $x = 0.5, t = 0$ with a velocity of -1, e.g., $(0.45 - 0.50)/(0.05 - 0) = -1$.

```
0.00   0.50    1.00000
0.05   0.45    0.99997
0.10   0.40    0.99984
0.15   0.35    0.99962
0.20   0.30    0.99939
```

The near constant value of the peak (maximum) is particularly noteworthy.

Figure 4.2 indicates that the gridding effect in Fig. 4.1 has been removed.

Also, the direction of the pulse can be reversed with cd=-cd added to pde1a (to produce the mirror image solution of Fig. 4.2).

Linear convection can now be introduced with $v \neq 0$, e.g., $v = 1$. The numerical and graphical output for alpha=1, nx=51 are shown in Table 4.3 and Fig. 4.3, respectively.

We can note the following details about the output shown in Table 4.3.

- The maximum value at $x = 0.5, t = 0$ is stationary in x, but decreases with t, e.g., $u(x = 0.5, t = 0) = 1.00000$, $u(x = 0.5, t = 0.2) = 0.74720$. That is, the fractional convection (with $\alpha = 1$) exactly balances the linear convection (with $v = 1$).

- The pulse maintains its symmetry around $x = 0.5$ exactly. For example

```
0.20   0.40    0.42678
0.20   0.50    0.74720
0.20   0.60    0.42678
```

Thus, the SPPDE is now essentially totally parabolic (even with $\alpha = 1$).

These features of the solution are confirmed in Fig. 4.3.

To conclude this discussion, the parameters v=1, cd=0 are used. With cd=0, the fractional derivative is deleted from eq. (4.1a) and only linear convection remains. The numerical output is shown in Table 4.4.

The graphical output is in Fig. 4.4.

Table 4.2: Numerical solution to eqs. (4.1), nx=101 (*Continues.*)

[1] 5

[1] 100

 alpha = 1.00 v = 0.00

```
     t      x     u(x,t)
   0.00   0.00   0.00000
   0.00   0.05   0.00000
   0.00   0.10   0.00000
            .
            .
            .
   0.00   0.40   0.36788
   0.00   0.45   0.77880
   0.00   0.50   1.00000
   0.00   0.55   0.77880
   0.00   0.60   0.36788
            .
            .
            .
   0.00   0.90   0.00000
   0.00   0.95   0.00000
   0.00   1.00   0.00000

   0.05   0.00   0.00000
   0.05   0.05   0.00000
   0.05   0.10   0.00001
            .
            .
            .
```

Table 4.2: (*Continued.*) Numerical solution to eqs. (4.1), nx=101 (*Continues.*)

0.05	0.35	0.36667
0.05	0.40	0.77558
0.05	0.45	0.99997
0.05	0.50	0.78204
0.05	0.55	0.36913
	.	
	.	
	.	
0.05	0.90	−0.00000
0.05	0.95	−0.00000
0.05	1.00	0.00000
0.10	0.00	0.00000
0.10	0.05	0.00001
0.10	0.10	0.00016
	.	
	.	
	.	
0.10	0.30	0.36551
0.10	0.35	0.77238
0.10	0.40	0.99984
0.10	0.45	0.78528
0.10	0.50	0.37043
	.	
	.	
	.	
0.10	0.90	−0.00000
0.10	0.95	−0.00000
0.10	1.00	0.00000

Table 4.2: (*Continued.*) Numerical solution to eqs. (4.1), nx=101

```
0.15   0.00     0.00000
0.15   0.05     0.00019
0.15   0.10     0.00240
         .
         .
         .
0.15   0.25     0.36440
0.15   0.30     0.76921
0.15   0.35     0.99962
0.15   0.40     0.78854
0.15   0.45     0.37179
         .
         .
         .
0.15   0.90     0.00000
0.15   0.95     0.00000
0.15   1.00     0.00000

0.20   0.00     0.00000
0.20   0.05     0.00262
0.20   0.10     0.02074
         .
         .
         .
0.20   0.20     0.36338
0.20   0.25     0.76612
0.20   0.30     0.99939
0.20   0.35     0.79186
0.20   0.40     0.37327
         .
         .
         .
0.20   0.90     0.00006
0.20   0.95     0.00006
0.20   1.00     0.00000

ncall = 646
```

Table 4.3: Numerical solution to eqs. (4.1), v=1, alpha=1, nx=51 (*Continues.*)

[1] 5

[1] 50

 alpha = 1.00 v = 1.00

```
     t      x     u(x,t)
  0.00   0.00   0.00000
  0.00   0.10   0.00000
            .
            .
            .
  0.00   0.40   0.36788
  0.00   0.50   1.00000
  0.00   0.60   0.36788
            .
            .
            .
  0.00   0.90   0.00000
  0.00   1.00   0.00000

  0.05   0.00   0.00000
  0.05   0.10   0.00000
            .
            .
            .
  0.05   0.40   0.39596
  0.05   0.50   0.91414
  0.05   0.60   0.39596
            .
            .
            .
  0.05   0.90   0.00000
  0.05   1.00   0.00000
```

Table 4.3: (*Continued.*) Numerical solution to eqs. (4.1), v=1, alpha=1, nx=51 (*Continues.*)

```
0.10   0.00    0.00000
0.10   0.10    0.00001
         .
         .
         .
0.10   0.40    0.41274
0.10   0.50    0.84688
0.10   0.60    0.41274
         .
         .
         .
0.10   0.90    0.00001
0.10   1.00    0.00000

0.15   0.00    0.00000
0.15   0.10    0.00004
         .
         .
         .
0.15   0.40    0.42218
0.15   0.50    0.79243
0.15   0.60    0.42218
         .
         .
         .
0.15   0.90    0.00004
0.15   1.00    0.00000

0.20   0.00    0.00000
0.20   0.10    0.00011
```

Table 4.3: (*Continued.*) Numerical solution to eqs. (4.1), v=1, alpha=1, nx=51

```
            .
            .
            .
  0.20  0.40    0.42678
  0.20  0.50    0.74720
  0.20  0.60    0.42678

            .
            .
            .

  0.20  0.90    0.00011
  0.20  1.00    0.00000


  ncall = 274
```

This output demonstrates the well-known numerical diffusion of a two point upwind FD as programmed in pde1a (with convection left to right).

```
#
# Array for ux
  ux=rep(0,nx);
  for(j in 1:nx){
    if(j==1){ux[1]=(u[1]-g_0(t))/dx};
    if(j>1) {ux[j]=(u[j]-u[j-1])/dx};
  }
```

4.3   SUMMARY AND CONCLUSIONS

Equation (4.1a) is readily accommodated within the MOL framework, and gives some unexpected solutions.

1. For v=0, alpha=1, eq. (4.1a) is totally convective (first order hyperbolic). That is, it has the characteristics of the linear advection eqs. (2.5). The validity of this conclusion is demonstrated in Fig. 4.2.

2. For v=1, alpha=1, the linear and fractional convection are exactly offset (these convection terms are equal in magnitude and opposite in sign) so that eq. (4.1a) is totally diffusive (parabolic). That is, it has the characteristics of the linear diffusion equation $\dfrac{\partial u}{\partial t} = \dfrac{\partial^2 u}{\partial x^2}$.

Table 4.4: Numerical solution to eqs. (4.1), v=1, cd=0, nx=51 (*Continues.*)

```
[1] 5

[1] 50

 alpha =  1.00  v =  1.00

      t      x     u(x,t)
   0.00   0.00   0.00000
   0.00   0.10   0.00000
   0.00   0.20   0.00012
   0.00   0.30   0.01832
   0.00   0.40   0.36788
   0.00   0.50   1.00000
   0.00   0.60   0.36788
   0.00   0.70   0.01832
   0.00   0.80   0.00012
   0.00   0.90   0.00000
   0.00   1.00   0.00000
                .
                .
                .
 Output for t = 0.05,
  0.1, 0.15 removed
                .
                .
                .
```

Table 4.4: (*Continued.*) Numerical solution to eqs. (4.1), v=1, cd=0, nx=51

```
0.20   0.00    0.00000
0.20   0.10    0.00000
0.20   0.20    0.00000
0.20   0.30    0.00004
0.20   0.40    0.00345
0.20   0.50    0.07642
0.20   0.60    0.44103
0.20   0.70    0.74583
0.20   0.80    0.41443
0.20   0.90    0.08415
0.20   1.00    0.00000


ncall = 343
```

3. To confirm the first conclusion, with v=1,cd=0 (so that the fractional derivative in eq. (4.1a) is deleted), the resulting solution with two point upwind MOL approximation of eq. (4.1a) has the well-known numerical diffusion (dispersion) as indicated by Fig. 4.4.

In conclusion, SFPDE (4.1a) has some unusual properties (hyperbolic and parabolic) that are not shared by the corresponding individual integer PDEs. As a word of caution, if α is increased from 1, the solution to eq. (4.1a) is more dispersed (diffuse) and can reach the boundaries in x, at which point it is no longer valid (BCs (4.1c,d) cannot be maintained). When this happens, the interval in x can be extended so that the nonzero part of the solution does not reach the boundaries. For example, the following extension in x accommodates the solution for $1 \leq \alpha \leq 2$

```
#
# Spatial grid
  xl=-2;xu=2;nx=51;dx=(xu-xl)/(nx-1);
  xj=seq(from=xl,to=xu,by=dx);
  cd=dx^(-alpha)/gamma(4-alpha);
```

Experimentation with α and v is left as an exercise for the reader.

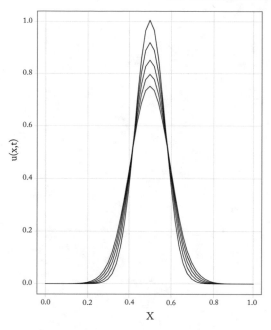

Figure 4.3: Numerical solution of eqs. (4.1), `v=1`, `alpha=1`, `nx=51`.

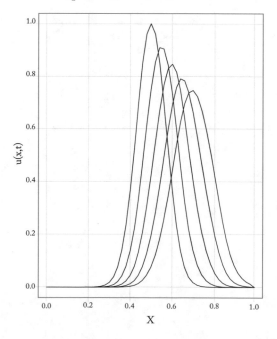

Figure 4.4: Numerical solution of eqs. (4.1), `v=1`, `cd=0`, `nx=51`.

CHAPTER 5

Nonlinear SFPDEs

5.1 INTRODUCTION

The example applications in the preceding chapters pertained to linear space fractional partial differential equations (SFPDEs). The following example applications pertain to nonlinear SF-PDEs. The first example has an analytical (exact) solution that can be used to compute the errors in the numerical solution for Dirichlet, Neumann and Robin boundary conditions (BCs).

5.1.1 EXAMPLE 1

The SFPDE system, including the exact solution, follows [1].

$$\frac{\partial u}{\partial x} = d(x,t)\frac{\partial^\alpha u}{\partial x^\alpha} + x^2 - 2u^2 \tag{5.1a}$$

with

$$d(x,t) = \Gamma(3-\alpha)x^{2+\alpha}t \tag{5.1b}$$

Equation (5.1a) has an inhomogeneous source term, x^2, and a nonlinear source term, $-2u^2$.
The initial condition (IC) for eq. (5.1a) is

$$u(x, t = 0) = 0; \; x_l = 0 \le x \le x_u = 1 \tag{5.1c}$$

The BCs for eq. (5.1a) are Dirichlet

$$u(x = x_l, t) = 0; \; u(x = x_u, t) = t \tag{5.1d,e}$$

The exact solution to eqs. (5.1) is

$$u_a(x,t) = x^2 t \tag{5.2}$$

The R routines for eqs. (5.1) and (5.2) follow.

5.1.2   MAIN PROGRAM

Listing 5.1: Main program for eqs. (5.1), (5.2)

```
#
# Nonlinear SFPDE
#
#   ut=d(x)*(d^alpha u/dx^alpha)+x^2-2*u^2
#
#   d(x)=gamma(3-alpha)*t*(x^(2+alpha))
#
#   xl < x < xu, 0 < t < tf, xl=0, xu=1
#
#   u(x,t=0)=0
#
#   c1(t)*ux(x=xl,t)+c2(t)*u(x=xl,t)=0
#
#   c3(t)*ux(x=xu,t)+c4(t)*u(x=xu,t)=t
#
#   ua(x,t)=(x^2)*t
#
# Delete previous workspaces
  rm(list=ls(all=TRUE))
#
# Access functions for numerical solution
  library("deSolve");
  setwd("f:/fractional/sfpde/chap5/dirichlet");
  source("pde1a.R");
#
# Parameters
  alpha=1.5;
#
# d(x)
  d=function(x,t) gamma(3-alpha)*(x^(2+alpha))*t;
#
# Analytical solution
  ua=function(x,t) (x^2)*t;
#
# Initial condition function (IC)
  f=function(x) ua(x,0);
```

```
#
# Boundary condition functions
  g_0=function(t) 0;
  g_L=function(t) t;
#
# Boundary condition coefficients
  c_1=function(t) 1;
  c_2=function(t) 0;
  c_3=function(t) 1;
  c_4=function(t) 0;
#
# Spatial grid
  xl=0;xu=1;nx=11;dx=(xu-xl)/(nx-1);
  xj=seq(from=xl,to=xu,by=dx);
  cd=dx^(-alpha)/gamma(4-alpha);
#
# Independent variable for ODE integration
  t0=0;tf=10;nt=6;dt=(tf-t0)/(nt-1);
  tout=seq(from=t0,to=tf,by=dt);
#
# a_jk coefficients
  A = matrix(0,nrow=nx-2,ncol=nx-1);
  for(j in 1:(nx-2)){
    for(k in 0:j){
    if (k==0){
      A[j,k+1]=(j-1)^(3-alpha)-j^(2-alpha)*(j-3+alpha);
    } else if (1 <= k && k<=j-1){
      A[j,k+1]=(j-k+1)^(3-alpha)-2*(j-k)^(3-alpha)+(j-k-1)^(3-
          alpha);
    } else
      A[j,k+1]=1;
    }
  }
#
# Initial condition
  u0=rep(0,nx-2);
  for(j in 1:(nx-2)){
    u0[j]=f(xj[j+1]);}
  ncall=0;
```

```
#
# ODE integration
  out=lsode(y=u0,times=tout,func=pde1a,
      rtol=1e-6,atol=1e-6,maxord=5);
  nrow(out)
  ncol(out)
#
# Allocate array for u(x,t)
  u=matrix(0,nt,nx);
#
# u(x,t), x ne xl,xu
  for(i in 1:nt){
    for(j in 2:(nx-1)){
      u[i,j]=out[i,j];
    }
  }
#
# Reset boundary values
  for(i in 1:nt){
   u[i,1]=ua(xl,tout[i]);
  u[i,nx]=ua(xu,tout[i]);
  }
#
# Numerical, analytical solutions, difference
  uap=matrix(0,nt,nx);
  for(i in 1:nt){
    for(j in 1:nx){
      uap[i,j]=ua((j-1)*dx,(i-1)*dt);
    }
  max_err=max(abs(u-uap));
  }
#
# Tabular numerical, analytical solutions,
# difference
  cat(sprintf("\n      t      x      u(x,t)    ua(x,t)         diff"))
    ;
  for(i in 1:nt){
  iv=seq(from=1,to=nx,by=1);
  for(j in iv){
```

```
      cat(sprintf("\n %6.2f%6.2f%10.5f%10.5f%12.3e",
        tout[i],xj[j],u[i,j],uap[i,j],u[i,j]-uap[i,j]));
    }
    cat(sprintf("\n"));
    }
#
# Plot numerical, analytical solutions
  matplot(xj,t(u),type="l",lwd=2,col="black",lty=1,
    xlab="x",ylab="u(x,t)",main="");
  matpoints(xj,t(uap),pch="o",col="black");
#
# Display maximum error
  cat(sprintf("  maximum error = %6.2e \n",max_err));
#
# Plot error at t = tf
  err_1=abs(u[nt,]-ua(xj[1:nx],tf));
  plot(xj,err_1,type="l",xlab="x",
      ylab="Max Error at t = tf",
      main="",col="black")
#
# Calls to ODE routine
  cat(sprintf("\n\n  ncall = %3d\n",ncall));
```

The main programs in Listings 3.1 and 5.1 are similar, but the details are repeated here for the purpose of a complete discussion of the Chapter 5 examples.

• Brief comments defining the test problem are followed by the deletion of previous files.

```
#
# Nonlinear SFPDE
#
#   ut=d(x)*(d^alpha u/dx^alpha)+x^2-2*u^2
#
#   d(x)=gamma(3-alpha)*t*(x^(2+alpha))
#
#   xl < x < xu, 0 < t < tf, xl=0, xu=1
#
#   u(x,t=0)=0
#
#   c1(t)*ux(x=xl,t)+c2(t)*u(x=xl,t)=0
#
```

```
#     c3(t)*ux(x=xu,t)+c4(t)*u(x=xu,t)=t
#
#     ua(x,t)=(x^2)*t
#
# Delete previous workspaces
  rm(list=ls(all=TRUE))
```

- The ODE integrator library deSolve is accessed. Note that the setwd (set working directory) uses / rather than the usual \.

```
#
# Access functions for numerical solution
  library("deSolve");
  setwd("f:/fractional/sfpde/chap5/dirichlet");
  source("pde1a.R");
```

The ODE/MOL routine is pde1a discussed subsequently.

- The order of the fractional derivative in eq. (5.1a) is specified.

```
#
# Parameters
  alpha=1.5;
```

- The function for the analytical solution eq. (5.2) is defined.

```
#
# Analytical solution
  ua=function(x,t) (x^2)*t;
```

- IC (5.1c) is the analytical solution at $t = 0$.

```
#
# Initial condition function (IC)
  f=function(x) ua(x,0);
```

- Functions for BCs (5.1d,e) are defined from the analytical solution of eq. (5.2) at $x = x_l, x_u$.

```
#
# Boundary condition functions
  g_0=function(t) 0;
  g_L=function(t) t;
```

- The BCs of eqs. (5.1d,e) are defined by specific values of $c_1(t), c_2(t), c_3(t), c_4(t)$ in eqs. (5.3c)–(5.3d).

```
#
# Boundary condition coefficients
  c_1=function(t) 1;
  c_2=function(t) 0;
  c_3=function(t) 1;
  c_4=function(t) 0;
```

- A spatial grid of 11 points is defined for $x_l = 0 \le x \le x_u = 1$, so that xj=0,1/10,...,1.

```
#
# Spatial grid
  xl=0;xu=1;nx=11;dx=(xu-xl)/(nx-1);
  xj=seq(from=xl,to=xu,by=dx);
  cd=dx^(-alpha)/gamma(4-alpha);
```

cd is a coefficient in eq. (5.1a) that is used in the ODE routine pde1a discussed subsequently.

- An interval in t of 6 points is defined for $0 \le t \le 10$ so that tout=0,2,...,10.

```
#
# Independent variable for ODE integration
  t0=0;tf=10;nt=6;dt=(tf-t0)/(nt-1);
  tout=seq(from=t0,to=tf,by=dt);
```

- The coefficients $a_{j,k}$ of eq. (1.2g) are defined with a series of ifs.

```
#
# a_jk coefficients
  A=matrix(0,nrow=nx-2,ncol=nx-1);
  for(j in 1:(nx-2)){
    for(k in 0:j){
    if (k==0){
      A[j,k+1]=(j-1)^(3-alpha)-j^(2-alpha)*(j-3+alpha);
    } else if (1 <= k && k<=j-1){
      A[j,k+1]=(j-k+1)^(3-alpha)-2*(j-k)^(3-alpha)+(j-k-1)^(3-alpha);
    } else
      A[j,k+1]=1;
    }
  }
```

- IC (5.1c) is defined.

```
#
# Initial condition
  u0=rep(0,nx-2);
  for(j in 1:(nx-2)){
    u0[j]=f(xj[j+1]);}
  ncall=0;
```

The counter for the calls to the ODE/MOL routine pde1a is also initialized.

With nx-2=11-2=9, the ICs are specified for the 9 ODEs at the interiors points in x, and do not include the boundary points $x = 0, 1$ since for the latter, the dependent variables $u(x = 0, t), u(x = 1, t)$ are defined by Dirichlet BCs (5.1d,e) and not by ODEs.

The coding pertaining to nx was selected since nx is also used in the ODE/MOL routine pde1a discussed subsequently to size vectors and control fors. As explained in the subsequent discussion, the value of nx passed to pde1a is the number of ODEs, 9 (the number of ICs) plus 2 (i.e., 11).

- The system of 9 MOL/ODEs is integrated by the library integrator lsode (available in deSolve). As expected, the inputs to lsode are the ODE function, pde1a, the IC vector u0, and the vector of output values of t, tout. The length of u0 (e.g., 9) informs lsode how many ODEs are to be integrated. func,y,times are reserved names.

```
#
# ODE integration
  out=lsode(y=u0,times=tout,func=pde1a,
      rtol=1e-6,atol=1e-6,maxord=5);
  nrow(out)
  ncol(out)
```

The numerical solution to the ODEs is returned in matrix out. In this case, out has the dimensions $nout \times (nx + 1) = 6 \times 9 + 1 = 10$, which are confirmed by the output from nrow(out),ncol(out) (included in the numerical output considered subsequently).

The offset $nx + 1$ is required since the first element of each column has the output t (also in tout), and the $2, ..., nx + 1 = 2, ..., 10$ column elements have the 9 ODE solutions.

- The solutions of the 9 ODEs returned in out by lsode are placed in u.

```
#
# Allocate array for u(x,t)
```

```
  u=matrix(0,nt,nx);
#
# u(x,t), x ne xl,xu
  for(i in 1:nt){
    for(j in 2:(nx-1)){
      u[i,j]=out[i,j];
    }
  }
```

Note that j=2,3,...,10 corresponding to the solution of the 9 ODEs.

- BCs (5.1d,e) are used to set $u(x = 0, t), u(x = 1, t)$ using the analytical solution of eq. (5.2).

```
#
# Reset boundary values
  for(i in 1:nt){
   u[i,1]=ua(xl,tout[i]);
   u[i,nx]=ua(xu,tout[i]);
   }
```

- The analytical $u(x, t)$ is used to compute the error in the numerical solution.

```
#
# Numerical, analytical solutions, difference
  uap=matrix(0,nt,nx);
  for(i in 1:nt){
    for(j in 1:nx){
      uap[i,j]=ua((j-1)*dx,(i-1)*dt);
    }
  max_err=max(abs(u-uap));
  }
```

The absolute maximum error is determined by the abs and max utilities.

- The numerical and analytical solutions, and the error are displayed.

```
#
# Tabular numerical, analytical solutions,
# difference
  cat(sprintf("\n      t      x      u(x,t)    ua(x,t)         diff"));
```

```
    for(i in 1:nt){
    iv=seq(from=1,to=nx,by=1);
    for(j in iv){
      cat(sprintf("\n %6.2f%6.2f%10.5f%10.5f%12.3e",
        tout[i],xj[j],u[i,j],uap[i,j],u[i,j]-uap[i,j]));
    }
    cat(sprintf("\n"));
    }
```

`iv=seq(from=1,to=nx,by=1)` is used to vary the number of output lines in x, in this case, every line with by=1.

- The numerical solution is plotted with lines (`matplot`) and the analytical solution is superimposed as points (`matpoints`).

```
#
# Plot numerical, analytical solutions
  matplot(xj,t(u),type="l",lwd=2,col="black",lty=1,
    xlab="x",ylab="u(x,t)",main="");
  matpoints(xj,t(uap),pch="o",col="black");
```

- The maximum absolute error along the solution is displayed.

```
#
# Display maximum error
  cat(sprintf("  maximum error = %6.2e \n",max_err));
```

- The absolute error at $t = t_f$ is plotted as a function of x.

```
#
# Plot error at t = tf
  err_1=abs(u[nt,]-ua(xj[1:nx],tf));
  plot(xj,err_1,type="l",xlab="x",
       ylab="Max Error at t = tf",
       main="",col="black")
```

- The numbers of calls to pde1a is displayed as a measure of the computational effort required to compute the solution.

```
#
# Calls to ODE routine
  cat(sprintf("\n\n  ncall = %3d\n",ncall));
```

The MOL/ODE routine, pde1a called by lsode, is discussed next.

5.1.3   SUBORDINATE ODE/MOL ROUTINE

Listing 5.2: ODE/MOL routine pde1a for eqs. (5.1), (5.2)

```
  pde1a=function(t,u,parms){
#
# Function pde1a computes the derivative
# vector of the ODEs approximating the
# PDE
#
# Allocate the vector of the ODE
# derivatives
  nx=nx-2;
  ut=rep(0,nx);
#
# Boundary approximations of uxx
  uxx=NULL;
#
# x=0
  u0=(2*dx*g_0(t)-c_2(t)*(4*u[1]-u[2]))/
     (2*dx*c_1(t)-3*c_2(t));
  uxx_0=2*u0-5*u[1]+4*u[2]-u[3];
  uxx[1]=u[2]-2*u[1]+u0;
#
# x=1
  un=(2*dx*g_L(t)+c_4(t)*(4*u[nx]-u[nx-1]))/
     (2*dx*c_3(t)+3*c_4(t));
  uxx[nx]=un-2*u[nx]+u[nx-1];
#
# Interior approximation of uxx
  for(k in 2:(nx-1)){
    uxx[k]=u[k+1]-2*u[k]+u[k-1];
  }
#
# PDE
#
# Step through ODEs
  for(j in 1:nx){
#
#   First term in series approximation of
```

```
#     fractional derivative
      ut[j]=A[j,1]*uxx_0;
#
#     Subsequent terms in series approximation
#     of fractional derivative
      for(k in 1:j){
        ut[j]=ut[j]+A[j,k+1]*uxx[k];
#
#     Next k (next term in series)
      }
      ut[j]=cd*d(xj[j+1],t)*ut[j]+(xj[j+1])^2-2*(u[j])^2;
#
# Next j (next ODE)
    }
#
# Increment calls to pde1a
  ncall <<- ncall+1;
#
# Return derivative vector of ODEs
  return(list(c(ut)));
  }
```

pde1a in Listings 3.2 and 5.2 are similar, but most of the details are repeated here for the purpose of a complete discussion of the Chapter 5 examples.

• The function is defined.

```
    pde1a=function(t,u,parms){
#
# Function pde1a computes the derivative
# vector of the ODEs approximating the
# PDE
```

t is the current value of t in eq. (5.1a). u is the 9-vector of ODE/MOL dependent variables. parm is an argument to pass parameters to pde1a (unused, but required in the argument list). The arguments must be listed in the order stated to properly interface with lsode called in the main program of Listing 5.1. The derivative vector of the LHS of eq. (5.1a) is calculated next and returned to lsode.

• An array, ut, is allocated for the derivative $\frac{\partial u}{\partial t}$ of eq. (5.1a).

```
#
```

```
# Allocate the vector of the ODE
# derivatives
  nx=nx-2;
  ut=rep(0,nx);
```

The value of nx is first reduced to the number of ODEs, that is, 11 to 9.

- The integer derivative $\dfrac{\partial^2 u}{\partial x^2}$ = uxx for use in eq. (1.1) ($n = 2$) is computed.

```
#
# Boundary approximations of uxx
  uxx=NULL;
#
# x=0
  u0=(2*dx*g_0(t)-c_2(t)*(4*u[1]-u[2]))/
      (2*dx*c_1(t)-3*c_2(t));
  uxx_0=2*u0-5*u[1]+4*u[2]-u[3];
  uxx[1]=u[2]-2*u[1]+u0;
#
# x=1
  un=(2*dx*g_L(t)+c_4(t)*(4*u[nx]-u[nx-1]))/
      (2*dx*c_3(t)+3*c_4(t));
  uxx[nx]=un-2*u[nx]+u[nx-1];
#
# Interior approximation of uxx
  for(k in 2:(nx-1)){
    uxx[k]=u[k+1]-2*u[k]+u[k-1];
  }
```

The details of this code are given in the discussion of Listing 3.2

- The series approximation of eqs. (1.2i), (1.2j) is computed.

```
#
# PDE
#
# Step through ODEs
  for(j in 1:nx){
#
#   First term in series approximation of
#   fractional derivative
```

```
      ut[j]=A[j,1]*uxx_0;
#
#     Subsequent terms in series approximation
#     of fractional derivative
      for(k in 1:j){
         ut[j]=ut[j]+A[j,k+1]*uxx[k];
#
#     Next k (next term in series)
      }
```

- The MOL approximation of eq. (5.1a) is programmed as

```
      ut[j]=cd*d(xj[j+1],t)*ut[j]+(xj[j+1])^2-2*(u[j])^2;
```

cd is a constant defined in the main program of Listing 5.1 that includes the factor $\frac{1}{\Delta x^2}$ required in the preceding FD approximations of the second derivative uxx. The programming of the inhomogeneous source term x^2 and the nonlinear source term $-2u^2$ in eq. (5.1a) is noteworthy.

- After all of the MOL/ODEs are programmed (all j), the counter ncall is incremented and returned to the main program of Listing 5.1. To conclude pde1a, the MOL derivative vector ut is returned to lsode as a list for the next step along the solution in t.

```
#
# Next j (next ODE)
   }
#
# Increment calls to pde1a
   ncall <<- ncall+1;
#
# Return derivative vector of ODEs
   return(list(c(ut)));
   }
```

This completes the programming of eqs. (5.1), (5.2). The output from the two routines of Listings 5.1, 5.2 is considered next.

5.1.4   MODEL OUTPUT

Abbreviated numerical output for eqs. (5.1), (5.2) is shown in Table 5.1.

We can observe the following details about the output in Table 5.1.

Table 5.1: Numerical solution to eqs. (5.1), (5.2), Dirichlet BCs (*Continues.*)

[1] 6

[1] 10

t	x	u(x,t)	ua(x,t)	diff
0.00	0.00	0.00000	0.00000	0.000e+00
0.00	0.10	0.00000	0.00000	0.000e+00
0.00	0.20	0.00000	0.00000	0.000e+00
0.00	0.30	0.00000	0.00000	0.000e+00
0.00	0.40	0.00000	0.00000	0.000e+00
0.00	0.50	0.00000	0.00000	0.000e+00
0.00	0.60	0.00000	0.00000	0.000e+00
0.00	0.70	0.00000	0.00000	0.000e+00
0.00	0.80	0.00000	0.00000	0.000e+00
0.00	0.90	0.00000	0.00000	0.000e+00
0.00	1.00	0.00000	0.00000	0.000e+00
2.00	0.00	0.00000	0.00000	0.000e+00
2.00	0.10	0.02000	0.02000	0.000e+00
2.00	0.20	0.08000	0.08000	0.000e+00
2.00	0.30	0.18000	0.18000	0.000e+00
2.00	0.40	0.32000	0.32000	0.000e+00
2.00	0.50	0.50000	0.50000	1.110e-16
2.00	0.60	0.72000	0.72000	-1.110e-16
2.00	0.70	0.98000	0.98000	-2.220e-16
2.00	0.80	1.28000	1.28000	-2.220e-16
2.00	0.90	1.62000	1.62000	-2.220e-16
2.00	1.00	2.00000	2.00000	0.000e+00

.
.
.

Table 5.1: (*Continued.*) Numerical solution to eqs. (5.1), (5.2), Dirichlet BCs

```
          Output for t = 4 to 8 removed
            .                     .
            .                     .
            .                     .
 10.00  0.00    0.00000    0.00000    0.000e+00
 10.00  0.10    0.10000    0.10000    2.776e-17
 10.00  0.20    0.40000    0.40000   -3.886e-16
 10.00  0.30    0.90000    0.90000   -2.554e-15
 10.00  0.40    1.60000    1.60000    1.887e-14
 10.00  0.50    2.50000    2.50000    4.219e-14
 10.00  0.60    3.60000    3.60000    1.821e-14
 10.00  0.70    4.90000    4.90000    4.885e-14
 10.00  0.80    6.40000    6.40000   -5.418e-14
 10.00  0.90    8.10000    8.10000    1.243e-14
 10.00  1.00   10.00000   10.00000    0.000e+00

maximum error = 5.42e-14

ncall =   27
```

- The solution array out is $6 \times 9 + 1 = 10$ as explained previously.

  [1] 6

  [1] 10

- The numerical and analytical solutions are the same for the IC at $t = 0$ since both are defined by the analytical solution of eq. (5.2).

- The output is for $x = 0, 0.1, ..., 1$ as programmed in Listing 5.1.

- The analytical and numerical solutions are the same at the boundaries $x = x_l = 0, x_u = 1$ since the analytical solution (eq. (5.2)) is used to set the numerical boundary values (Listings 5.1, 5.2).

- The output is for $t = 0, 2, ..., 10$ as programmed in Listing 5.1.

- The maximum error is 5.42e-14. Therefore, the error is small, even with a spatial grid of only 11 points. This is a result of a smooth solution as reflected in eq. (5.2)

• The computational effort is small, `ncall = 27`, again reflecting the smooth solution.

The graphical output is in Figs. 5.1, 5.2.

This completes the discussion of the MOL solution of eqs. (5.1). Since the analytical solution, eq. (5.2), is not a function of α, the numerical solution will be essentially the same for all α (the errors in Fig. 5.2 will be different, but small, which the reader can easily verify by changing α in Listing 5.1).

Also, the use of Neumann and Robin BCs follows in a similar way by using the analytical solution eq. (5.2) to define the boundary conditions. The only required changes are in the coefficients of BCs (5.3c), (5.3d).

Dirichlet BCs (Listing 5.1)

```
#
# Boundary condition coefficients
  c_1=function(t) 1;
  c_2=function(t) 0;
  c_3=function(t) 1;
  c_4=function(t) 0;
```

Neumann BCs

```
#
# Boundary condition coefficients
  c_1=function(t) 0;
  c_2=function(t) 1;
  c_3=function(t) 0;
  c_4=function(t) 1;
```

Robin BCs

```
#
# Boundary condition coefficients
  c_1=function(t) -1;
  c_2=function(t)  1;
  c_3=function(t)  1;
  c_4=function(t)  1;
```

These coefficients are discussed further in Chapter 3, Example 2.

The discussion of a second example of a nonlinear SFPDE is next.

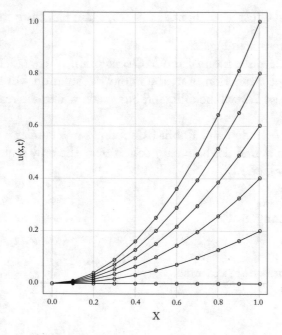

Figure 5.1: Numerical, analytical solutions of eqs. (5.1), (5.2), Dirichlet BCs, $\alpha = 1.5$ lines - num, points - anal.

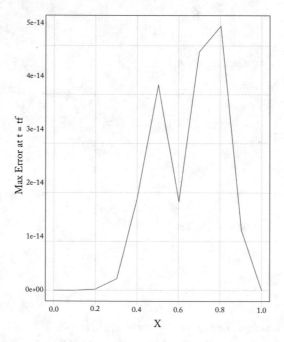

Figure 5.2: Error in the numerical solution of eqs. (5.1), (5.2).

5.2   EXAMPLE 2

This example is based on the fractional Fisher-Kolmogorov equation which is the fractional diffusion equation with a nonlinear logistic source term.

$$\frac{\partial u}{\partial t} = \frac{\partial^{\alpha} u}{\partial x^{\alpha}} + u(1 - u^q); \; x_l = -5 \le x \le x_u = 10 \qquad (5.3a)$$

where q is an arbitrary (specified) constant.

The initial condition (IC) for eq. (5.3a) is

$$u(x, t = 0) = 1/(1 + ae^{bx})^s \qquad (5.3b)$$

and the boundary conditions (BCs) are

$$c_2(t)\frac{\partial u(x = x_l, t)}{\partial x} + c_1(t)u(x = x_l, t) = g_0(t) \qquad (5.3c)$$

$$c_4(t)\frac{\partial u(x = x_u, t)}{\partial x} + c_3(t)u(x = x_u, t) = g_L(t) \qquad (5.3d)$$

where $c_1(t) = c_3(t) = 1, c_2(t) = c_4(t) = 0$ (Dirichlet BCs) and

$$g_0(t) = 1/(1 + ae^{b(x_l - ct)})^s \qquad (5.3e)$$

$$g_L(t) = 1/(1 + ae^{b(x_u - ct)})^s \qquad (5.3f)$$

Equations (5.3b)–(5.3f) are special cases of the analytical solution for $\alpha = 2$ (for the integer Fisher-Kolmogorov equation [2], p305).

$$u(x, t) = 1/(1 + ae^{b(x - ct)})^s \qquad (5.4a)$$

Equation (5.4a) is used to test the following routines (for $\alpha = 2$).

a, b, c, s are constants defined as

$$a = \sqrt{2} - 1; \; b = \frac{q}{[2(q + 2)]^{1/2}}; \; c = \frac{(q + 4)}{[2(q + 2)]^{1/2}}; \; s = \frac{2}{q} \qquad (5.4b)$$

A main program for eqs. (5.3), (5.4) is listed next.

5.2.1   MAIN PROGRAM

Listing 5.3: Main program for eqs. eqs. (5.3), (5.4)

```
#
# Space fractional Fisher-Kolmogorov
#
```

```
#    ut=(d^alpha u/dx^alpha)+u*(1-u^q)
#
#    xl < x < xu, 0 < t < tf, xl=-5, xu=10
#
#    u(x,t=0)=1/(1+a*e^(b*x))^s
#
#    c1(t)*ux(x=xl,t)+c2(t)*u(x=xl,t)=g_0(t)
#
#    c3(t)*ux(x=xu,t)+c4(t)*u(x=xu,t)=g_L(t)
#
#    g_0(t)=1/(1+a*exp(b*(-5-c*t)))^s
#
#    g_L(t)=1/(1+a*exp(b*(10-c*t)))^s
#
#    a=2^(0.5)-1; b=q/(2*(q+2))^(0.5);
#
#    c=(q+4)/(2*(q+2))^(0.5); s=2/q;
#
#    alpha=2:
#
#       ua(x,t) = 1/(1+a*exp(b*(x-c*t)))^s
#
# Delete previous workspaces
  rm(list=ls(all=TRUE))
#
# Access functions for numerical solution
  library("deSolve");
  setwd("f:/fractional/sfpde/chap5/dirichlet");
  source("pde1b.R");source("ua.R");
#
# Parameters
  ncase=1;
  if(ncase==1){alpha=2};
  if(ncase==2){alpha=1.5};
  q=1;a=2^(0.5)-1;
  b=q/(2*(q+2))^(0.5);
  c=(q+4)/(2*(q+2))^(0.5);
  s=2/q;
#
```

```
# Boundary condition functions
  g_0=function(t) ua(xl,t);
  g_L=function(t) ua(xu,t);
#
# Boundary condition coefficients
  c_1=function(t) 1;
  c_2=function(t) 0;
  c_3=function(t) 1;
  c_4=function(t) 0;
#
# Spatial grid
  xl=-5;xu=10;nx=31;dx=(xu-xl)/(nx-1);
  xj=seq(from=xl,to=xu,by=dx);
  cd=dx^(-alpha)/gamma(4-alpha);
#
# Independent variable for ODE integration
  t0=0;tf=5;nt=6;dt=(tf-t0)/(nt-1);
  tout=seq(from=t0,to=tf,by=dt);
#
# a_jk coefficients
  A=matrix(0,nrow=nx-2,ncol=nx-1);
  for(j in 1:(nx-2)){
    for(k in 0:j){
    if (k==0){
      A[j,k+1]=(j-1)^(3-alpha)-j^(2-alpha)*(j-3+alpha);
    } else if (1 <= k && k<=j-1){
      A[j,k+1]=(j-k+1)^(3-alpha)-2*(j-k)^(3-alpha)+(j-k-1)^(3-
          alpha);
    } else
      A[j,k+1]=1;
    }
  }
#
# Initial condition
  u0=rep(0,nx-2);
  for(j in 1:(nx-2)){
    u0[j]=ua(xj[j+1],t0);}
  ncall=0;
#
```

```
# ODE integration
  out=lsode(y=u0,times=tout,func=pde1b,
      rtol=1e-6,atol=1e-6,maxord=5);
  nrow(out)
  ncol(out)
#
# Allocate array for u(x,t)
  u=matrix(0,nt,nx);
#
# u(x,t), x ne xl,xu
  for(i in 1:nt){
    for(j in 2:(nx-1)){
      u[i,j]=out[i,j];
    }
  }
#
# Reset boundary values
  for(i in 1:nt){
   u[i,1]=g_0(tout[i]);
  u[i,nx]=g_L(tout[i]);
  }
#
# Analytical solution
  if(ncase==1){
    uap=matrix(0,nt,nx);
    for(i in 1:nt){
      for(j in 1:nx){
        uap[i,j]=ua(xl+(j-1)*dx,(i-1)*dt);
      }
    }
#
#   Tabular numerical, analytical solutions,
#   difference
    cat(sprintf("\n        t      x     u(x,t)
                   ua(x,t)         diff"));
    for(i in 1:nt){
    iv=seq(from=1,to=nx,by=5);
    for(j in iv){
      cat(sprintf("\n %6.2f%6.2f%10.5f%10.5f%12.3e",
```

```
          tout[i],xj[j],u[i,j],uap[i,j],u[i,j]-uap[i,j]));
      }
        cat(sprintf("\n"));
      }
#
#   Display maximum error
    max_err=max(abs(u-uap));
    cat(sprintf("\n    maximum error = %6.2e \n",max_err));
#
#   Plot numerical, analytical solutions
    matplot(xj,t(u),type="l",lwd=2,col="black",lty=1,
      xlab="x",ylab="u(x,t)",main="");
    matpoints(xj,t(uap),pch="o",col="black");
#
# Plot error at t = tf
  err_1=abs(u[nt,]-ua(xj[1:nx],tf));
  plot(xj,err_1,type="l",xlab="x",
      ylab="Max Error at t = tf",
      main="",col="black")
  }
#
# Numerical solution
  if(ncase==2){
#
#   Tabular numerical solution
#   difference
    cat(sprintf("\n        t      x     u(x,t)"));
    for(i in 1:nt){
    iv=seq(from=1,to=nx,by=5);
    for(j in iv){
      cat(sprintf("\n %6.2f%6.2f%10.5f",
      tout[i],xj[j],u[i,j]));
    }
      cat(sprintf("\n"));
    }
#
#   Plot numerical, analytical solutions
    matplot(xj,t(u),type="l",lwd=2,col="black",lty=1,
      xlab="x",ylab="u(x,t)",main="");
```

```
    }
#
# Calls to ODE routine
  cat(sprintf("\n\n   ncall = %3d\n",ncall));
```

The main programs in Listings 5.1, 5.3 are similar so only the differences are considered here.

- Brief comments defining the test problem are followed by the deletion of previous files.

```
#
# Space fractional Fisher-Kolmogorov
#
#   ut=(d^alpha u/dx^alpha)+u*(1-u^q)
#
#   xl < x < xu, 0 < t < tf, xl=-5, xu=10
#
#   u(x,t=0)=1/(1+a*e^(b*x))^s
#
#   c1(t)*ux(x=xl,t)+c2(t)*u(x=xl,t)=g_0(t)
#
#   c3(t)*ux(x=xu,t)+c4(t)*u(x=xu,t)=g_L(t)
#
#   g_0(t)=1/(1+a*exp(b*(-5-c*t)))^s
#
#   g_L(t)=1/(1+a*exp(b*(10-c*t)))^s
#
#   a=2^(0.5)-1; b=q/(2*(q+2))^(0.5);
#
#   c=(q+4)/(2*(q+2))^(0.5); s=2/q;
#
#   alpha=2:
#
#     ua(x,t) = 1/(1+a*exp(b*(x-c*t)))^s
#
# Delete previous workspaces
  rm(list=ls(all=TRUE))
```

- The ODE integrator library deSolve is accessed.

```
#
```

```
# Access functions for numerical solution
  library("deSolve");
  setwd("f:/fractional/sfpde/chap5/dirichlet");
  source("pde1b.R");source("ua.R");
```

The ODE/MOL routine is pde1b discussed subsequently. ua.R is a routine for the analytical solution of eqs. (5.4).

- Two cases are programmed: ncase=1 corresponds to $\alpha = 2$ for which the analytical solution to the integer Fisher-Kolmogorov equation, eq. (5.4a), is used to verify the numerical solution, and ncase=2 for $\alpha = 1.5$ for which an analytical solution is not readily available. The constants q, a, b, c, s are defined numerically according to eqs. (5.4b).

```
#
# Parameters
  ncase=1;
  if(ncase==1){alpha=2};
  if(ncase==2){alpha=1.5};
  q=1;a=2^(0.5)-1;
  b=q/(2*(q+2))^(0.5);
  c=(q+4)/(2*(q+2))^(0.5);
  s=2/q;
```

- BC functions (5.3c), (5.3d) are defined by the analytical solution of eq. (5.4a) with $x = x_l, x_u$.

```
#
# Boundary condition functions
  g_0=function(t) ua(xl,t);
  g_L=function(t) ua(xu,t);
```

- The coefficients in BCs (5.3c), (5.3d) are defined for Dirichlet BCs.

```
#
# Boundary condition coefficients
  c_1=function(t) 1;
  c_2=function(t) 0;
  c_3=function(t) 1;
  c_4=function(t) 0;
```

- A spatial grid with 31 points is defined for $-5 \le x \le 10$ so that xj=-5,-4.5,..,10.

```
#
# Spatial grid
  xl=-5;xu=10;nx=31;dx=(xu-xl)/(nx-1);
  xj=seq(from=xl,to=xu,by=dx);
  cd=dx^(-alpha)/gamma(4-alpha);
```

The asymmetry of the grid around $x = 0$ was used to accommodate the traveling wave solution of eq. (5.4a) which is a function of the Lagrangian variable $x - ct$ with $c > 0$ (the solution travels left to right toward $x = x_u$).

• An interval in t, $0 \leq t \leq 5$, is defined with 6 output points so tout=0,1,...,5.

```
#
# Independent variable for ODE integration
  t0=0;tf=5;nt=6;dt=(tf-t0)/(nt-1);
  tout=seq(from=t0,to=tf,by=dt);
```

• IC (5.3b) is defined with the analytical solution of eqs. (5.4) and $t = 0$.

```
#
# Initial condition
  u0=rep(0,nx-2);
  for(j in 1:(nx-2)){
    u0[j]=ua(xj[j+1],t0);}
  ncall=0;
```

• The $31 - 2 = 29$ MOL/ODEs at the interior points in x are integrated. The ODE function pde1b is discussed subsequently. The arguments of lsode are described after Listing 5.1.

```
#
# ODE integration
  out=lsode(y=u0,times=tout,func=pde1b,
      rtol=1e-6,atol=1e-6,maxord=5);
  nrow(out)
  ncol(out)
```

The solution matrix out is $6 \times 29 + 1 = 30$ as explained for Listing 5.1 and confirmed by the numerical output that follows.

• The BC values $u(x = x_l, t)$, $u(x = x_u, t)$ are reset since they are not returned by pde1b in the solution matrix out. The subscripts 1,nx correspond to the boundary values of x.

```
#
# Reset boundary values
  for(i in 1:nt){
   u[i,1]=g_0(tout[i]);
   u[i,nx]=g_L(tout[i]);
   }
```

• For ncase=1, the analytical solution of eq. (5.4a) is placed in uap for plotting.

```
#
# Analytical solution
  if(ncase==1){
    uap=matrix(0,nt,nx);
    for(i in 1:nt){
      for(j in 1:nx){
        uap[i,j]=ua(xl+(j-1)*dx,(i-1)*dt);
      }
    }
#
#   Tabular numerical, analytical solutions,
#   difference
    cat(sprintf("\n       t       x       u(x,t)
                   ua(x,t)          diff"));
    for(i in 1:nt){
    iv=seq(from=1,to=nx,by=5);
    for(j in iv){
      cat(sprintf("\n %6.2f%6.2f%10.5f%10.5f%12.3e",
      tout[i],xj[j],u[i,j],uap[i,j],u[i,j]-uap[i,j]));
    }
      cat(sprintf("\n"));
    }
```

The numerical and analytical sloution, and the difference are displayed. Every fifth x is selected with by=5.

• After $t = t_f = 5$, the maximum error is determined and displayed with the max, abs utilities.

```
#
#   Display maximum error
    max_err=max(abs(u-uap));
    cat(sprintf("\n   maximum error = %6.2e \n",max_err));
```

- The numerical and analytical solutions are superimposed with `matplot` and `matpoints`.

```
#
#    Plot numerical, analytical solutions
     matplot(xj,t(u),type="l",lwd=2,col="black",lty=1,
       xlab="x",ylab="u(x,t)",main="");
     matpoints(xj,t(uap),pch="o",col="black");
   }
```

Two transposes, `t(u)`,`t(uap)`, are required so that the number of rows of the transposed matrices equals the number of elements of `xj` (31). The numerical and analytical solutions are then plotted parametrically in t. The final } terminates `ncase=1`.

- The maximum error $t = t_f = 5$ is plotted agaist x.

```
#
# Plot error at t = tf
  err_1=abs(u[nt,]-ua(xj[1:nx],tf));
  plot(xj,err_1,type="l",xlab="x",
      ylab="Max Error at t = tf",
      main="",col="black")
```

- For `ncase=2`, an analytical solution is not available, so the preceding code is repeated without ua.

- The number of calls to `pde1b` at the conclusion of the solutions is displayed.

```
#
# Calls to ODE routine
   cat(sprintf("\n\n  ncall = %3d\n",ncall));
```

The subordinate ODE/MOL routine, `pde1b`, called by `lsode` is considered next.

5.2.2   SUBORDINATE ODE/MOL ROUTINE

Listing 5.4: ODE/MOL routine pd1a for eqs. (5.3), (5.4)

```
  pde1b=function(t,u,parms){
#
# Function pde1b computes the derivative
# vector of the ODEs approximating the
# PDE
```

```
#
# Allocate the vector of the ODE
# derivatives
  nx=nx-2;
  ut=rep(0,nx);
#
# Boundary approximations of uxx
  uxx=NULL;
#
# x=0
  u0=(2*dx*g_0(t)-c_2(t)*(4*u[1]-u[2]))/
     (2*dx*c_1(t)-3*c_2(t));
  uxx_0=2*u0-5*u[1]+4*u[2]-u[3];
  uxx[1]=u[2]-2*u[1]+u0;
#
# x=1
  un=(2*dx*g_L(t)+c_4(t)*(4*u[nx]-u[nx-1]))/
     (2*dx*c_3(t)+3*c_4(t));
  uxx[nx]=un-2*u[nx]+u[nx-1];
#
# Interior approximation of uxx
  for(k in 2:(nx-1)){
    uxx[k]=u[k+1]-2*u[k]+u[k-1];
  }
#
# PDE
#
# Step through ODEs
  for(j in 1:nx){
#
#   First term in series approximation of
#   fractional derivative
    ut[j]=A[j,1]*uxx_0;
#
#   Subsequent terms in series approximation
#   of fractional derivative
    for(k in 1:j){
      ut[j]=ut[j]+A[j,k+1]*uxx[k];
#
```

```
#    Next k (next term in series)
     }
     ut[j]=cd*ut[j]+u[j]*(1-u[j]^q);
#
# Next j (next ODE)
     }
#
# Increment calls to pde1b
     ncall <<- ncall+1;
#
# Return derivative vector of ODEs
     return(list(c(ut)));
     }
```

pde1b is similar to pde1a in Listing 5.2, so only the differences are considered.

- The function is defined.

```
   pde1b=function(t,u,parms){
```

- Equation (5.3a) is programmed.

```
    ut[j]=cd*ut[j]+u[j]*(1-u[j]^q);
```

The straightforward coding of the nonlinear logistic source function in eq. (5.3a) is clear.

The output from the R routines in Listings 5.3 and 5.4 is considered next.

5.2.3   MODEL OUTPUT

Abbreviated numerical output for eqs. (5.3), (5.4), ncase=1, is shown in Table 5.2.

We can observe the following details about the output in Table 5.2.

- The solution array out is $6 \times 29 + 1 = 30$ as explained previously.

  [1] 6

  [1] 30

- The numerical and analytical solutions are the same for the IC at $t = 0$ since both are defined by the analytical solution of eqs. (5.4).

- The output is for $x = -5, -2.5, ..., 10$ as programmed in Listing 5.3, with every fifth value displayed.

Table 5.2: Numerical solution to eqs. (5.3), (5.4), Dirichlet BCs, ncase=1 (*Continues.*)

[1] 6

[1] 30

t	x	u(x,t)	ua(x,t)	diff
0.00	-5.00	0.90051	0.90051	0.000e+00
0.00	-2.50	0.75710	0.75710	0.000e+00
0.00	0.00	0.50000	0.50000	0.000e+00
0.00	2.50	0.21645	0.21645	0.000e+00
0.00	5.00	0.05697	0.05697	0.000e+00
0.00	7.50	0.01031	0.01031	0.000e+00
0.00	10.00	0.00153	0.00153	0.000e+00
1.00	-5.00	0.95483	0.95483	0.000e+00
1.00	-2.50	0.88189	0.88187	2.330e-05
1.00	0.00	0.71823	0.71816	6.989e-05
1.00	2.50	0.44466	0.44472	-6.713e-05
1.00	5.00	0.17551	0.17563	-1.173e-04
1.00	7.50	0.04261	0.04257	3.980e-05
1.00	10.00	0.00734	0.00734	0.000e+00

```
        .                       .
        .                       .
        .                       .
    Output for t = 2 to 4 removed
        .                       .
        .                       .
        .                       .
```

Table 5.2: (*Continued.*) Numerical solution to eqs. (5.3), (5.4), Dirichlet BCs, ncase=1

```
5.00 -5.00    0.99833    0.99833    0.000e+00
5.00 -2.50    0.99538    0.99539   -3.473e-06
5.00  0.00    0.98727    0.98728   -5.517e-06
5.00  2.50    0.96528    0.96529   -7.813e-06
5.00  5.00    0.90797    0.90798   -1.347e-05
5.00  7.50    0.77321    0.77324   -3.005e-05
5.00 10.00    0.52451    0.52451    0.000e+00

maximum error = 2.49e-04

ncall = 198
```

- The analytical and numerical solutions are the same at the boundaries $x = x_l = -5, x_u = 10$ since the analytical solution (eq. (5.4a)) is used to set the numerical boundary values (Listings 5.3, 5.4).

- The output is for $t = 0, 1, ..., 5$ as programmed in Listing 5.3.

- The maximum error is 2.49e-04.

- The computational effort is modest ncall = 198.

The graphical output is in Figs. 5.3, 5.4.

The changing values of the solution at $x = -5, 10$ according to eqs. (5.3c) to (5.3f) is clear. The graphical output of Figs. 5.3, 5.4 reflects the numerical output in Table 5.2.

Abbreviated numerical output for ncase=2 (in Listing 5.3) is shown in Table 5.3.

The computational effort is modest, ncall = 226. The graphical output is in Fig. 5.5.

The solution for ncase=2, $\alpha = 1.5$ is similar to the solution of Fig. 5.3 ($\alpha = 2$) for the same BCs, eqs. (5.3c) to (5.3f). That is, eq. (5.1a) interpolates in essentially the same way at the interior points for $\alpha = 1.5, 2$.

However, this is not the case for smaller α. As $\alpha \to 1$, eq. (5.3a) approaches the first order advection equation with the nonlinear logistic source term (see also eqs. (2.5)). The SFPDE is therefore strongly parabolic (approaching the second order diffusion equation) for $\alpha \to 2$ with smoothing from diffusion, and strongly hyperbolic for $\alpha \to 1$ with front propagation from convection.

For the latter case, as the solution moves left to right with increasing t, the boundary at $x = 10$ causes numerical inaccuracy. Generally, boundary effects for hyperbolic PDEs have been studied extensively and are not considered further here other than to point out that the spurious boundary effects could possibly be reduced by increasing the number of points in x, and

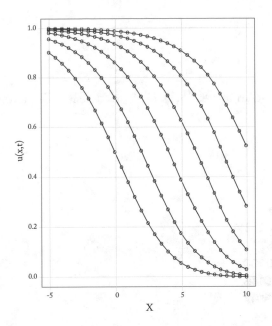

Figure 5.3: Numerical, analytical solutions of eqs. (5.3), (5.4), ncase=1 lines - num, points - anal.

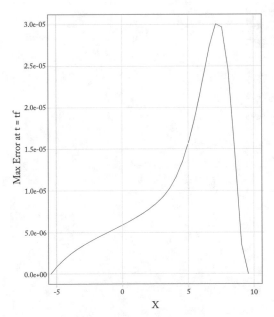

Figure 5.4: Error in the numerical solution of eqs. (5.3), (5.4), ncase=1.

Table 5.3: Numerical solution to eqs. (5.3), (5.4), Dirichlet BCs, ncase=2 (*Continues.*)

```
[1] 6

[1] 30

     t      x      u(x,t)
  0.00 -5.00    0.90051
  0.00 -2.50    0.75710
  0.00  0.00    0.50000
  0.00  2.50    0.21645
  0.00  5.00    0.05697
  0.00  7.50    0.01031
  0.00 10.00    0.00153

  1.00 -5.00    0.95483
  1.00 -2.50    0.87386
  1.00  0.00    0.69489
  1.00  2.50    0.41672
  1.00  5.00    0.17794
  1.00  7.50    0.06764
  1.00 10.00    0.00734
    .            .
    .            .
    .            .

Output for t = 2 to
     4 removed
    .            .
    .            .
    .            .
```

Table 5.4: (*Continued.*) Numerical solution to eqs. (5.3), (5.4), Dirichlet BCs, ncase=2

```
5.00 -5.00    0.99833
5.00 -2.50    0.99319
5.00  0.00    0.98030
5.00  2.50    0.95073
5.00  5.00    0.88695
5.00  7.50    0.73936
5.00 10.00    0.52451

ncall = 226
```

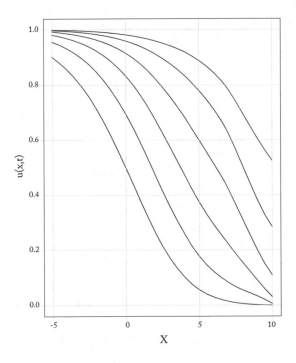

Figure 5.5: Numerical solution of eqs. (5.3), (5.4), ncase=2.

by extending the right boundary beyond $x = 10$. But ultimately, sharp fronts from hyperbolic PDEs approaching exit boundaries will cause difficulties without the use of special numerical methods.

5.3 SUMMARY AND CONCLUSIONS

The preceding examples demonstrate the MOL solution of a SFPDE with a nonlinear source term. In particular, the numerical MOL approach for including the nonlinear source term is straightforward, and the solutions are computed with acceptable accuracy and computational effort provided the solution is sufficiently smooth.

In the next chapter (Chapter 6, vol. 2), we consider simultaneous (coupled) SFPDEs.

REFERENCES

[1] Salehi, Y. (2017), Private communication, July, 2017. 129

[2] Schiesser, W.E. (2017), *Spline Collocation Methods for Partial Differential Equations*, Wiley, Hoboken, NJ. 147

APPENDIX A

Analytical Caputo Differentiation of Selected Functions

This appendix is an introduction to the analytical (exact) fractional differentiation of selected functions according to the following definition of the Caputo derivative

$$D^\alpha f(x) = \frac{\partial^\alpha f(x)}{\partial x^\alpha} = \frac{1}{\Gamma(n-\alpha)} \int_a^x \frac{d^n f(s)/ds^n}{(x-s)^{\alpha+1-n}} ds; \ n-1 < \alpha < n \qquad (A.1)$$

where n is the smallest integer greater than α. For $\alpha = n$, eq. (A.1) gives the integer derivative

$$\frac{\partial^\alpha f(x)}{\partial x^\alpha} = \frac{d^n f(x)}{dx^n}; \ \alpha = n \qquad (A.2)$$

(see eq. (A.5) as an example).

An important special case is eq. (A.1) applied to $f(x) = C$ (C a constant). With $a = 0, \alpha < 1, n = 1$,

$$D^\alpha C = \frac{\partial^\alpha C}{\partial x^\alpha} = \frac{1}{\Gamma(n-\alpha)} \int_a^x \frac{0}{(x-s)^{\alpha+1-n}} ds = 0; \qquad (A.3a)$$

(from $d^n f(s)/ds^n = d^n C/ds^n = 0$), that is, the Caputo derivative of a constant is zero. Generally, this is not the case for alternate defintions of fractional derivatives, e.g., the Riemann-Liouville derivative.

As another example of the use of eq. (A.1), $f(x) = x, a = 0, \alpha = 1/2, n = 1$ ([1], p14)

$$D^{1/2} x = \frac{\partial^{1/2} x}{\partial x^{1/2}}$$

$$= \frac{-1}{\Gamma(1/2)} \int_0^x \frac{1}{(x-s)^{1/2}} d(x-s)$$

$$= \frac{-1}{\sqrt{\pi}} \int_{\sqrt{x}}^0 \frac{1}{u} du^2$$

$$= \frac{1}{\sqrt{\pi}} \int_0^{\sqrt{x}} \frac{2u}{u} du$$

$$= \frac{2}{\sqrt{\pi}}(\sqrt{x} - 0)$$

or

$$D^{1/2}x = \frac{2\sqrt{x}}{\sqrt{\pi}} \tag{A.3b}$$

Generally useful values of the Gamma function are listed below (from [1, 2]).

Table A.1: Selected values of the Gamma function

x	$\Gamma(x)$
1/2	$\sqrt{\pi}$
1, 2	1
3/2	$(1/2)\sqrt{\pi}$
n	$(n-1)!;\ n \geq 1$ n integer; $0! = 1$
n	$\pm\infty$ poles $n = 0, -1, -2, \ldots$
$z + 1$	$z\Gamma(z)$

Selected Caputo derivatives (from eq. (A.1)) follow, first for the power function x^β (Table A.2) [1].

Power function	Caputo derivative
$D^\alpha x^\beta$	$\dfrac{\Gamma(\beta + 1)}{\Gamma(\beta - \alpha + 1)}x^{\beta-\alpha}$
Special case: $D^{1/2}x$	$\dfrac{\Gamma(2)}{\Gamma(3/2)}x^{1/2} =$
(eq. (A.3b))	$\dfrac{1}{(1/2)\Gamma(1/2)}x^{1/2} = \dfrac{2}{\sqrt{\pi}}x^{1/2}$

Table A.2: Caputo derivatives for the function x^2

$$D^\alpha x^2 \qquad \frac{2}{\Gamma(3-\alpha)}x^{2-\alpha}$$

$$\alpha = 1/3;\ D^{1/3}x^2 \qquad \frac{\Gamma(3)}{\Gamma(3-1/3)}x^{2-1/3} =$$

$$\frac{2}{\Gamma(8/3)}x^{5/3} \approx 1.33x^{5/3}$$

$$\alpha = 1/2;\ D^{1/2}x^2 \qquad \frac{\Gamma(3)}{\Gamma(3-1/2)}x^{2-1/2} =$$

$$\frac{8}{3\sqrt{\pi}}x^{3/2} \approx 1.5x^{3/2}$$

$$\alpha = 3/4;\ D^{3/4}x^2 \qquad \frac{\Gamma(3)}{\Gamma(3-3/4)}x^{2-3/4} =$$

$$\frac{2}{\Gamma(9/4)}x^{5/4} \approx 1.77x^{5/4}$$

An interesting special case from [1] p32 (with $\Gamma(0) = \pm\infty$), $\Gamma(1) = \Gamma(2) = 1, \Gamma(3) = 2$)

$$D^\alpha x^2 = \frac{2}{\Gamma(3-\alpha)}x^{2-\alpha} = \begin{cases} x^2, & \alpha \to 0 \\ 2x, & \alpha \to 1 \\ 2, & \alpha \to 2 \\ 0, & \alpha \to 3 \end{cases} \qquad (A.5)$$

demonstrates that the Caputo derivative: (1) reduces to the integer derivative for integer α (see eq. (A.2)), and (2) can be used for interpolation for fractional α.

With the fractional derivative of the power function available, the fractional derivative of other functions can be stated in terms of a Taylor series. For example, the fractional exponential function is[1]

$$D^\alpha e^{\lambda x} = \sum_{k=0}^{\infty} \frac{\lambda^{k+n} x^{k+n-\alpha}}{\Gamma(k+1+n-\alpha)} \qquad (A.6a)$$

Then the fractional derivative of $\sin(x)$ is

$$D^\alpha \sin(x) = D^\alpha \left(\frac{e^{i\lambda x} - e^{-i\lambda x}}{2i}\right) = \sum_{k=0}^{\infty} \frac{(i\lambda)^{k+n} - (-i\lambda)^{k+n}}{2i} \frac{x^{k+n-\alpha}}{\Gamma(k+1+n-\alpha)} \qquad (A.6b)$$

The fractional derivative of $\cos(x)$ is

$$D^\alpha \cos(x) = D^\alpha \left(\frac{e^{i\lambda x} + e^{-i\lambda x}}{2} \right) = \sum_{k=0}^{\infty} \frac{(i\lambda)^{k+n} + (-i\lambda)^{k+n}}{2} \frac{x^{k+n-\alpha}}{\Gamma(k+1+n-\alpha)} \quad \text{(A.6c)}$$

Equations (A.6) can be used for the numerical computation of the Caputo derivative of $e^{\lambda x}, \sin(x), \cos(x)$ as a function of x, α.

    An extensive literature pertaining to the Caputo fractional derivative is available which the reader/analyst can consult for mathematical details and applications. Here, selected fractional derivatives are presented that might be useful in testing numerical methods for fractional ordinary and partial differential equations.

REFERENCES

[1] Ishteva, M.K. (2005), *Properies and Applications of the Caputo Fractional Operator*, M.S. Thesis, Department of Mathematics, Universitaet Karlsruhe. 165, 166, 167

[2] Particular values of the Gamma function are given in `https://en.wikipedia.org/wik i/Particular_values_of_the_Gamma_function` 166

---

[1]The Caputo derivative/integral of eq. (A.1) can be stated for various functions in terms of the Mittag-Leffler (ML) two-parameter functions, $E_{\alpha,\beta}(x)$, defined as

$$E_{\alpha,\beta}(x) = \sum_{k=0}^{\infty} \frac{x^k}{\Gamma(\alpha k + \beta)}; \ \alpha, \beta > 0 \quad \text{(A.4)}$$

Further consideration of the ML functions is not given here.

APPENDIX B

Derivation of a SFPDE Analytical Solution

B.1 INTRODUCTION

This appendix details the derivation and verification of the analytical (exact) solution of a SFPDE. The SFPDE and the analytical solution can then be used as a test problem for the numerical algorithm of eqs. (1.2a)–(1.2j).

B.2 SFPDE EQUATIONS

The SFPDE is a variation of eq. (3.1) [2]

$$\frac{\partial u(x,t)}{\partial t} = \frac{24x^{\alpha}}{\Gamma(5+\alpha)}\frac{\partial^{\alpha}u(x,t)}{\partial x^{\alpha}} - 2u(x,t) \tag{B.1}$$

The initial condition (IC) for eq. (B.1) is

$$u(x, t = 0) = x^{4+\alpha} \tag{B.2a}$$

and the boundary conditions (BCs) are

$$u(x = 0, t) = 0; \; u(x = 1, t) = e^{-t} \tag{B.2b,c}$$

The analytical solution of eqs. (B.1), (B.2) is

$$u(x,t) = e^{-t}x^{4+\alpha} \tag{B.3}$$

which is confirmed by substitution in eqs. (B.1), (B.2).

To verify that $u(x,t)$ of eq. (B.3) is a solution to eqs. (B.1), (B.2), we first substitute into the definition of eq. (1.1)

$$\frac{\partial^{\alpha}u(x,t)}{\partial x^{\alpha}} = \frac{1}{\Gamma(2-\alpha)}\int_{0}^{x}(x-s)^{1-\alpha}\frac{\partial^2(e^{-t}s^{4+\alpha})}{\partial s^2}ds$$

$$= \frac{(4+\alpha)(3+\alpha)e^{-t}}{\Gamma(2-\alpha)}\int_{0}^{x}(x-s)^{1-\alpha}s^{2+\alpha}ds$$

$$= \frac{(4+\alpha)(3+\alpha)e^{-t}}{\Gamma(2-\alpha)} \int_0^x (x-s)^{\theta-1} s^{\beta} ds$$

where $\theta = 2 - \alpha$ and $\beta = 2 + \alpha$.

To evalute the integral

$$\int_0^x (x-s)^{\theta-1} s^{\beta} ds$$

the change of variable $\tau = s/x$ is used so that

$$s = x\tau; \quad ds = x d\tau$$

Then, from the Euler integral of the first kind (the Euler beta function) [1]

$$\int_0^x (x-s)^{\theta-1} s^{\beta} ds = \int_0^1 (x-x\tau)^{\theta-1} (x\tau)^{\beta} x d\tau = x^{\theta+\beta} \int_0^1 (1-\tau)^{\theta-1} \tau^{\beta} d\tau$$

$$= \frac{\Gamma(\theta)\Gamma(\beta+1)}{\Gamma(\theta+\beta+1)} x^{\theta+\beta} = \frac{\Gamma(2-\alpha)\,\Gamma(3+\alpha)}{\Gamma(2-\alpha+3+\alpha)} x^4$$

Therefore,

$$\frac{\partial^{\alpha} u(x,t)}{\partial x^{\alpha}} = \frac{(4+\alpha)(3+\alpha)e^{-t}}{\Gamma(2-\alpha)} \int_0^x (x-s)^{\theta-1} s^{\beta} ds$$

$$= \frac{(4+\alpha)(3+\alpha)e^{-t}}{\Gamma(2-\alpha)} \frac{\Gamma(2-\alpha)\Gamma(3+\alpha)}{\Gamma(5)} x^4 = \frac{(4+\alpha)(3+\alpha)\Gamma(3+\alpha)}{\Gamma(5)} e^{-t} x^4$$

$$= \frac{\Gamma(5+\alpha)}{24} e^{-t} x^4$$

Then, substitution in eq. (B.1) gives

$$-e^{-t} x^{4+\alpha} = \frac{24\, x^{\alpha}}{\Gamma(5+\alpha)} \frac{\Gamma(5+\alpha)\, e^{-t} x^4}{24} - 2e^{-t} x^{4+\alpha} = e^{-t} x^{4+\alpha} - 2e^{-t} x^{4+\alpha}$$

or

$$-e^{-t} x^{4+\alpha} = -e^{-t} x^{4+\alpha}$$

which confirms the solution of eq. (B.3).

To conclude this discussion, eqs. (B.1) to (B.3) are used to test the numerical algorithm of eqs. (1.2a)–(1.2j). A main program and the ODE/MOL routine follows.

B.2.1   MAIN PROGRAM

Listing B.1: Main program for eqs. (B.1), (B.2), (B.3)

```
#
# SFPDE
```

```
#
#    ut=d(x)*(d^alpha u/dx^alpha)-2*u(x,t)
#
#    xl < x < xu, 0 < t < tf, xl=0, xu=1
#
#    u(x,t=0)=x^(4+alpha)
#
#    u(x=xl,t)=0; u(x=xu,t)=exp(-t)
#
#    d(x)=24/gamma(5+alpha)*x^alpha
#
#    ua(x,t)=exp(-t)*x^(4+alpha)
#
# Delete previous workspaces
  rm(list=ls(all=TRUE))
#
# Access functions for numerical solution
  library("deSolve");
  setwd("f:/fractional/sfpde/appB");
  source("pde1a.R");
#
# Parameters
  for(ncase in 1:5){
    if(ncase==1){alpha=1;}
    if(ncase==2){alpha=1.25;}
    if(ncase==3){alpha=1.5;}
    if(ncase==4){alpha=1.75;}
    if(ncase==5){alpha=2;}
#
# d(x)
  d=function(x,t) 24/gamma(5+alpha)*x^alpha;
#
# Analytical solution
  ua=function(x,t) exp(-t)*x^(4+alpha);
#
# Initial condition function (IC)
  f=function(x) ua(x,0);
#
# Boundary condition functions (BCs)
```

```
    g_0=function(t) ua(xl,t);
    g_L=function(t) ua(xu,t);
#
# Spatial grid
    xl=0;xu=1;nx=41;dx=(xu-xl)/(nx-1);
    xj=seq(from=xl,to=xu,by=dx);
    cd=dx^(-alpha)/gamma(4-alpha);
#
# Independent variable for ODE integration
    t0=0;tf=1;nt=6;dt=(tf-t0)/(nt-1);
    tout=seq(from=t0,to=tf,by=dt);
    ncall=0;
#
# a_jk coefficients
    A=matrix(0,nrow=nx-2,ncol=nx-1);
    for(j in 1:(nx-2)){
      for(k in 0:j){
      if (k==0){
        A[j,k+1]=(j-1)^(3-alpha)-j^(2-alpha)*(j-3+alpha);
      } else if (1 <= k && k<=j-1){
        A[j,k+1]=(j-k+1)^(3-alpha)-2*(j-k)^(3-alpha)+(j-k-1)^(3-
            alpha);
      } else
        A[j,k+1]=1;
      }
    }
#
# Initial condition
    nx=nx-2;
    u0=rep(0,nx);
    for(j in 1:nx){
      u0[j]=f(xj[j+1]);}
#
# ODE integration
    out=lsode(y=u0,times=tout,func=pde1a,
        rtol=1e-6,atol=1e-6,maxord=5);
    nrow(out)
    ncol(out)
#
```

```
# Allocate array for u(x,t)
  nx=nx+2;
  u=matrix(0,nt,nx);
#
# u(x,t), x ne xl,xu
  for(i in 1:nt){
    for(j in 2:(nx-1)){
      u[i,j]=out[i,j];
    }
  }
#
# Reset boundary values
  for(i in 1:nt){
   u[i,1]=g_0(tout[i]);
   u[i,nx]=g_L(tout[i]);
   }
#
# Numerical, analytical solutions, maximum difference
  uap=matrix(0,nt,nx);
  for(i in 1:nt){
    for(j in 1:nx){
      uap[i,j]=ua((j-1)*dx,(i-1)*dt);
    }
  max_err=max(abs(u-uap));
   }
#
# Tabular numerical, analytical solutions,
# difference
  cat(sprintf("\n\n   alpha = %4.2f\n",alpha));
  cat(sprintf("\n       t     x     u(x,t)    ua(x,t)          diff"))
    ;
  for(i in 1:nt){
  iv=seq(from=1,to=nx,by=4);
  for(j in iv){
    cat(sprintf("\n %6.2f%6.2f%10.5f%10.5f%12.3e",
      tout[i],xj[j],u[i,j],uap[i,j],u[i,j]-uap[i,j]));
  }
  cat(sprintf("\n"));
  }
```

```
#
# Plot numerical, analytical solutions
  matplot(xj,t(u),type="l",lwd=2,col="black",lty=1,
    xlab="x",ylab="u(x,t)",main="");
  matpoints(xj,t(uap),pch="o",col="black");
#
# Display maximum error
  cat(sprintf("\n   Maximum error = %6.2e \n",max_err));
#
# Plot error at t = tf
  err_1=abs(u[nt,]-ua(xj[1:nx],tf));
  plot(xj,err_1,type="l",xlab="x",
      ylab="Max Error at t = tf",
      main="",col="black")
#
# Calls to ODE routine
  cat(sprintf("\n   ncall = %3d\n",ncall));
#
# Next alpha (ncase)
  }
```

Listing B.1 is similar to Listing 2.1, so only the differences are discussed here.

- Brief comments state the SFPDE.

```
#
# SFPDE
#
#   ut=d(x)*(d^alpha u/dx^alpha)-2*u(x,t)
#
#   xl < x < xu, 0 < t < tf, xl=0, xu=1
#
#   u(x,t=0)=x^(4+alpha)
#
#   u(x=xl,t)=0; u(x=xu,t)=exp(-t)
#
#   d(x)=24/gamma(5+alpha)*x^alpha
#
#   ua(x,t)=exp(-t)*x^(4+alpha)
```

- The ODE/MOL is pde1a. This routine is discussed next.

```
#
# Access functions for numerical solution
  library("deSolve");
  setwd("f:/fractional/sfpde/appB");
  source("pde1a.R");
```

- The order of the fractional derivative in eq. (B.1) is varied through five values.

```
#
# Parameters
  for(ncase in 1:5){
    if(ncase==1){alpha=1;}
    if(ncase==2){alpha=1.25;}
    if(ncase==3){alpha=1.5;}
    if(ncase==4){alpha=1.75;}
    if(ncase==5){alpha=2;}
```

- The variable coefficient in eq. (B.1) is defined.

```
#
# d(x)
  d=function(x,t) 24/gamma(5+alpha)*x^alpha;
```

- The analytical solution of eq. (B.3) is defined.

```
#
# Analytical solution
  ua=function(x,t) exp(-t)*x^(4+alpha);
```

- IC (B.2a) is defined in terms of the analytical solution of eq. (B.3) with $t = t_0 = 0$.

```
#
# Initial condition function (IC)
  f=function(x) ua(x,0);
```

- BCs (B.2b,c) are defined as functions g_0, g_L by the analytical solution of eq. (B.3) with $x = x_l = 0, x = x_u = 1$ (see Listing 2.1 discussion).

- The spatial interval is $0 \leq x \leq 1$ with 41 grid points and the time scale is $t_0 = 0 \leq t \leq t_f = 1$ with 6 output points (see Listing 2.1 discussion).

- The coefficients A of the numerical algorithm of eqs. (1.2a)–(1.2j) are defined (see Listing 2.1 discussion).

- IC (B.2a) is defined with the analytical solution of eq. (B.3) (see Listing 2.1 discussion).

- The integration of the 41−2=39 ODEs at the interior points in x is with lsode and pde1a.

```
out=lsode(y=u0,times=tout,func=pde1a,
    rtol=1e-6,atol=1e-6,maxord=5);
```

This choice of an integrator (lsode) illustrates one possible selection from deSolve. lsodes is another possibility. In general, although lsodes is more complex internally than lsode (due to the sparse matrix processing), the complexity can substantially reduce the level of computation for large ODE systems. The reader can verify that the numerical solution with lsodes is the same as for lsode.

- The numerical and analytical solutions, and their differences, are computed and displayed as in Listing 2.1. This code ends with the display of the number of calls to pde1a and completion of the for in ncase (see Listing 2.1 discussion).

The ODE/MOL routine pde1a follows.

B.2.2   ODE/MOL ROUTINE

Listing B.2: ODE/MOL routines for eqs. (B.1), (B.2), (B.3)

```
  pde1a=function(t,u,parms){
#
# Function pde1a computes the derivative
# vector of the ODEs approximating the
# PDE
#
# Allocate the vector of the ODE
# derivatives
  ut=rep(0,nx);
#
# Boundary approximations of uxx
  uxx=NULL;
  uxx_0=2*g_0(t)-5*u[1]+4*u[2]-u[3];
  uxx[1]=u[2]-2*u[1]+g_0(t);
  uxx[nx]=g_L(t)-2*u[nx]+u[nx-1];
#
# Interior approximation of uxx
  for(k in 2:(nx-1)){
    uxx[k]=u[k+1]-2*u[k]+u[k-1];
```

```
   }
#
# PDE
#
# Step through ODEs
   for(j in 1:nx){
#
#   First term in series approximation of
#   fractional derivative
    ut[j]=uxx_0*A[j,1];
#
#   Subsequent terms in series approximation
#   of fractional derivative
    for(k in 1:j){
      ut[j]=ut[j]+uxx[k]*A[j,k+1];
#
#   Next k (next term in series)
    }
#
#   ODE with fractional derivative
    ut[j]=cd*d(xj[j+1],t)*ut[j]-2*u[j];
#
# Next j (next ODE)
   }
#
# Increment calls to pde1a
   ncall <<- ncall+1;
#
# Return derivative vector of ODEs
   return(list(c(ut)));
   }
```

Listing B.2 is the same as Listing 2.2 except for one line.

```
Listing (2.2)
#
#   ODE with fractional derivative
    ut[j]=cd*d(xj[j+1],t)*ut[j]+p(xj[j+1],t);

Listing (B.2)
```

```
#
#    ODE with fractional derivative
     ut[j]=cd*d(xj[j+1],t)*ut[j]-2*u[j];
```

The inhomogeneous source term, p(x,t), is used in eq. (2.1a) while the term −2u is used in eq. (B.1). In other words, for eq. (2.1a) an explicit function is used while in eq. (B.1) the numerical solution is used.

This may seem like an inconsequential difference, but actually it is important since it illustrates the MOL implementation of two different types of SFPDEs. A comparison of the numerical output follows.

B.2.3   NUMERICAL OUTPUT

Numerical output is shown in Table B.1.

Table B.1: Comparison of the numerical solution for eqs. (2.1), (2.2), (2.3), and eqs. (B.1), (B.2), $\alpha = 1$ (*Continues.*)

```
Listings 2.1, 2.2, Table 2.1

    alpha = 1.00

       t      x      u(x,t)     ua(x,t)         diff
    0.00   0.00    0.00000     0.00000     0.000e+00
    0.00   0.10    0.00001     0.00001     0.000e+00
    0.00   0.20    0.00032     0.00032     0.000e+00
    0.00   0.30    0.00243     0.00243     0.000e+00
    0.00   0.40    0.01024     0.01024     0.000e+00
    0.00   0.50    0.03125     0.03125     0.000e+00
    0.00   0.60    0.07776     0.07776     0.000e+00
    0.00   0.70    0.16807     0.16807     0.000e+00
    0.00   0.80    0.32768     0.32768     0.000e+00
    0.00   0.90    0.59049     0.59049     0.000e+00
    0.00   1.00    1.00000     1.00000     0.000e+00
```

Table B.1: (*Continued.*) Comparison of the numerical solution for eqs. (2.1), (2.2), (2.3), and eqs. (B.1), (B.2), $\alpha = 1$ (*Continues.*)

```
          .                    .
          .                    .
          .                    .
      Output for t = 0.2 to 0.8 removed
          .                    .
          .                    .
          .                    .
1.00  0.00   0.00000   0.00000   0.000e+00
1.00  0.10   0.00000   0.00000   8.397e-07
1.00  0.20   0.00013   0.00012   8.492e-06
1.00  0.30   0.00092   0.00089   2.975e-05
1.00  0.40   0.00384   0.00377   7.138e-05
1.00  0.50   0.01164   0.01150   1.401e-04
1.00  0.60   0.02885   0.02861   2.425e-04
1.00  0.70   0.06221   0.06183   3.850e-04
1.00  0.80   0.12109   0.12055   5.424e-04
1.00  0.90   0.21751   0.21723   2.783e-04
1.00  1.00   0.36788   0.36788   0.000e+00
```

Table B.1: (*Continued.*) Comparison of the numerical solution for eqs. (2.1), (2.2), (2.3), and eqs. (B.1), (B.2), $\alpha = 1$ (*Continues.*)

```
Maximum error = 5.48e-04

ncall = 222

Listings B.1, B.2, Table B.1

alpha = 1.00

    t     x     u(x,t)    ua(x,t)        diff
  0.00  0.00   0.00000    0.00000    0.000e+00
  0.00  0.10   0.00001    0.00001    0.000e+00
  0.00  0.20   0.00032    0.00032    0.000e+00
  0.00  0.30   0.00243    0.00243    0.000e+00
  0.00  0.40   0.01024    0.01024    0.000e+00
  0.00  0.50   0.03125    0.03125    0.000e+00
  0.00  0.60   0.07776    0.07776    0.000e+00
  0.00  0.70   0.16807    0.16807    0.000e+00
  0.00  0.80   0.32768    0.32768    0.000e+00
  0.00  0.90   0.59049    0.59049    0.000e+00
  0.00  1.00   1.00000    1.00000    0.000e+00
          .                  .
          .                  .
          .                  .

     Output for t = 0.2 to 0.8 removed
          .                  .
          .                  .
          .                  .
```

Table B.1: (*Continued.*) Comparison of the numerical solution for eqs. (2.1), (2.2), (2.3), and eqs. (B.1), (B.2), $\alpha = 1$

```
1.00   0.00    0.00000    0.00000    0.000e+00
1.00   0.10    0.00000    0.00000    2.699e-07
1.00   0.20    0.00012    0.00012    2.812e-06
1.00   0.30    0.00090    0.00089    9.890e-06
1.00   0.40    0.00379    0.00377    2.375e-05
1.00   0.50    0.01154    0.01150    4.661e-05
1.00   0.60    0.02869    0.02861    8.065e-05
1.00   0.70    0.06196    0.06183    1.280e-04
1.00   0.80    0.12073    0.12055    1.854e-04
1.00   0.90    0.21738    0.21723    1.555e-04
1.00   1.00    0.36788    0.36788    0.000e+00

Maximum error = 2.31e-04

ncall = 222
```

B.3   SUMMARY AND CONCLUSIONS

In this appendix the derivation and verification of an analytical (exact) solution of a SFPDE is illustrated. The numerical MOL solution of the SFPDE is also compared with the analytical solution. Thus, this appendix can be considered a self contained discussion of the analytical solution and the numerical solution (based on the algorithm of eqs. (1.2a)–(1.2j)) of a SFPDE.

REFERENCES

[1] Ishteva, M.K. (2005), *Properies and Applications of the Caputo Fractional Operator*, M.S. Thesis, Department of Mathematics, Universitaet Karlsruhe. 170

[2] Sousa, E. (2011), Numerical approximations for fractional diffusion equations via splines, *Computers and Mathematics with Applications*, **62**, 938–944. 169

Authors' Biographies

YOUNES SALEHI

My research focus is applied mathematics broadly. This includes numerical linear algebra, optimization and solving differential equations. My primary research interest concerns the areas of numerical analysis, scientific computing and high performance computing with particular emphasis on the numerical solution of ordinary differential equations (ODEs) and partial differential equations (PDEs).

One focus of my work is programming efficient numerical methods for ODEs and PDEs. I have extensive experience in MATLAB, Maple, Mathematica and R programming of transportable numerical method routines, but I am also experienced in programming in C, C++ and C#, and could readily apply these programming systems to numerical ODE/PDEs.

Recently, I have become interested in fractional differential equations (FDEs), especially the numerical solution of fractional initial value problems (FIVPs) and space fractional differential equations (SFPDEs).

WILLIAM E. SCHIESSER

William E. Schiesser is Emeritus McCann Professor of Computational Biomedical Engineering and Chemical and Biomolecular Engineering, and Professor of Mathematics at Lehigh University. His research is directed toward numerical methods and associated software for ordinary, differential-algebraic and partial differential equations (ODE/DAE/PDEs). He is the author, coauthor or coeditor of 18 books, and his ODE/DAE/PDE computer routines have been accessed by some 5,000 colleges and universities, corporations and government agencies.

Index